Studies in Systems, Decision and Control

Volume 44

Series editor

Janusz Kacprzyk, Polish Academy of Sciences, Warsaw, Poland
e-mail: kacprzyk@ibspan.waw.pl

About this Series

The series "Studies in Systems, Decision and Control" (SSDC) covers both new developments and advances, as well as the state of the art, in the various areas of broadly perceived systems, decision making and control- quickly, up to date and with a high quality. The intent is to cover the theory, applications, and perspectives on the state of the art and future developments relevant to systems, decision making, control, complex processes and related areas, as embedded in the fields of engineering, computer science, physics, economics, social and life sciences, as well as the paradigms and methodologies behind them. The series contains monographs, textbooks, lecture notes and edited volumes in systems, decision making and control spanning the areas of Cyber-Physical Systems, Autonomous Systems, Sensor Networks, Control Systems, Energy Systems, Automotive Systems, Biological Systems, Vehicular Networking and Connected Vehicles, Aerospace Systems, Automation, Manufacturing, Smart Grids, Nonlinear Systems, Power Systems, Robotics, Social Systems, Economic Systems and other. Of particular value to both the contributors and the readership are the short publication timeframe and the world-wide distribution and exposure which enable both a wide and rapid dissemination of research output.

More information about this series at http://www.springer.com/series/13304

Dirk Buchholz

Bin-Picking

New Approaches for a Classical Problem

 Springer

Dirk Buchholz
Braunschweig
Germany

ISSN 2198-4182   ISSN 2198-4190   (electronic)
Studies in Systems, Decision and Control
ISBN 978-3-319-26498-1   ISBN 978-3-319-26500-1   (eBook)
DOI 10.1007/978-3-319-26500-1

Library of Congress Control Number: 2015955371

Springer Cham Heidelberg New York Dordrecht London

Springer International Publishing AG Switzerland is part of Springer Science+Business Media (www.springer.com)

If every instrument, at command, or from
a preconception of its master's will, could
accomplish its work (as the story goes of the statues
of Daedalus; or what the poet tells us of the tripods
of Vulcan, "that they moved of their own accord
into the assembly of the gods"), the shuttle would
then weave, and the lyre play of itself; nor would
the architect want servants, or the master
slaves. [12]

Aristotle, 384–322 BCE

Acknowledgments

The present book was written in the context of my work as a Ph.D. student at the Institut für Robotik und Prozessinformatik of the Technische Universität Carolo Wilhelmina in Braunschweig. I would like to express my gratitude to my advisor Prof. Wahl for the continuous support of my Ph.D. study and research. I would also like to thank Prof. Henrich for being part of the thesis committee.

Further, I would like to thank my colleagues of the institute for assisting me with advises, help, and discussions during my work. With topics not only within the scope of research, all of you enriched the time of my studies. Especially, I would like to say thank you to Daniel, who contributed to this work in diverse ways, not only during our research time in Italy. Further thanks go to Simon, who took the time to do the first review of this work. My thanks also go to Natasha, for putting the finishing touches to this text.

Special thanks go to my students Marcus, Alex, Eike, and Andreas for sharing my research interests and for working on my ideas with lots of enthusiasm and great involvement. I also thank Antje for the cooperation.

My personal thanks go to my parents and my friends, for continuously supporting me and for backing me not only during the preparation of this book but always.

The most important person involved in this work was my wife Kathi.

Thank you Kathi for always sustaining me, even in difficult phases of my work. Thank you for always being there for me.

Braunschweig   Dirk Buchholz
December 2014

Acknowledgments

Contents

List of Figures

Abstract

The automation of handling tasks has been an important scientific topic since the development of the first industrial robots. The first step in the chain of scientific challenges to be solved is the automatic grasping of objects. One of the most famous examples in this context is the well-known "bin-picking" problem. To pick up objects scrambled in a box is an easy task for humans, but its automation is very complex. Besides the localization of the object, meaning the estimation of the object's pose (orientation and position), it has to be ensured that a collision-free path can be found to safely grasp the objects. For over 50 years, researchers have published approaches towards generic solutions to this problem, but unfortunately no industry-applicable, generic system has been developed yet.

In this thesis, three different approaches to solve the bin-picking problem are described. More precisely, different solutions to the pose estimation problem are introduced, each paired with additional functionalities to complete it for application in a bin-picking station. It is described, how modern sensors can be used for efficient bin-picking as well as how classic sensor concepts can be applied for novel bin-picking techniques. Three complete systems are described and compared.

First, 3D point clouds, generated using a laser scanner, are used as basis. Employing the known random sample matching algorithm and modifications of it, paired with a very efficient depth map based collision avoidance mechanism results in a very robust bin-picking approach.

In the second approach, all computations are done on depth maps. This allows the use of 2D image analysis techniques to fulfill the tasks and results in real-time data analysis. Combined with force/torque and acceleration sensors, a near-time optimal bin-picking system emerges.

As a third option, surface normal maps are employed as a basis for pose estimation. In contrast to known approaches, the normal maps are not used for 3D data computation but directly for the object localization problem. This enables the application of a new class of sensors for bin-picking.

All three methods are compared and advantages and disadvantages of each approach are discussed.

Chapter 1
Introduction—Automation and the Need for Pose Estimation

One definition of automation says: "*Automation is the transfer of human work to automatons, realized with the help of machines.*" [71]. And many other definitions are very similar to this. Taking this literally, it says that automation is the replacement of human workers by machines. The vision of machines releasing humans from hard work is far from being new. As quoted at the very beginning of this document, Aristotle already has had this vision of "*instruments that, at command, or from a preconception could accomplish the work of its master at its will*". But, although this wish is nearly as old as civilization itself, there are many workplaces in industrial manufacturing halls, where humans have not been replaced by machines, even when the tasks are not ergonomic, monotonic or very simple. Even allegedly easy tasks are not automated, yet. Suppose asking an infant child to pick its favorite toy out of a box. The kid will, of course, fulfill this task without thinking a lot and without injuring him or her self. The described task is not as easy as it is explained. Humans are only able to fulfill it easily because they can be regarded as very complex machines, equipped with a very big portfolio of actors and sensors. These sensors are for example a sophisticated stereo vision system and tactile sensors in the fingers and in every joint, without being complete. With these senses, combined with intelligence, a superior experience and internal and external model knowledge, it is easy for even young children to recognize single objects on a pile of different ones and to pick them up. Humans have uncountable built in sensors, databases, path planners and efficient collision avoidance mechanisms even including on-line trajectory generation.

The tasks described above are collectively known in scientific literature as the "*bin-picking problem*" and a lot of diverse and sincere challenges have to be faced to solve it. Literally since the first digital image and industrial robot, this problem is famous in science. Researchers have been proposing methods, related to the bin-picking problem for over 50 years. But, till today, no working, generic solution to the problem can be found in modern industry. The reason for this is that at least some of the complex abilities of human workers have to be rebuilt. More precisely, industrial robots have to be equipped with sensors and with some kind of artificial intelligence,

© Springer International Publishing Switzerland 2016
D. Buchholz, *Bin-Picking*, Studies in Systems, Decision and Control 44,
DOI 10.1007/978-3-319-26500-1_1

at least in the context of a specific task. But, the algorithms should not be too specific as special solutions always mean a big effort in updating them to new demands.

Getting to the point, in industry, parts, scrambled in boxes or on piles or lying on a table, have to be located, picked and placed in a defined way. Here, it is obvious that no presumptions can be made of the object's pose. Like mentioned above, the way to solve this is to enable the robot to acquire information about its workspace, interpret or "understand" this information and generate movements and actions dependent upon it. The bin-picking problem is only one of several tasks in which it is important to estimate the poses of distinct objects in a scene. Service robots can be mentioned as a second example and there are many more. A robot with the autonomy to understand its surroundings can be efficient, robust and in some cases only then applicable in industry, even when the autonomy only covers one simple task.

So, the main and basic problem that has to be solved in any autonomous robot system is the *analysis* of appropriate environmental data. In most cases, the best option is to rely on visual data as these can be obtained contactlessly. Here, the choice of a suitable vision sensor is only the first problem of many and cannot be solved in general. What can be said in general is that a 3D world has to be analyzed and understood to enable robots to solve their specific tasks autonomously. It is obvious that 3D sensor data would be best to describe a 3D scene. But, even with 3D sensors becoming cheaper in the last years, they are not in every case the best option. And, if 3D sensors are used and 3D data is available, this 3D data has to be analyzed efficiently. The vision system of a robot therefore consists of two main parts, namely the vision acquisition and its analysis. Both parts have to interact well to build an efficient entity.

Although robot vision is such an enormously important aspect for autonomous robots and for many decades, research has been conducted on this topic, no generic bin-picking approach has been developed that made its way into industrial production lines, yet. This thesis describes three novel approaches to solve the bin-picking problem by giving three new pose estimation techniques embedded into a working framework. Three more steps towards the goal of autonomous robots.

Chapter 2
Bin-Picking—5 Decades of Research

Reviewing the literature on the bin-picking problem leads to scientific papers written more than 50 years ago. Bin-picking is a meta discipline which combines several sub-disciplines, like scene analysis, object recognition, object localization (or pose estimation), grasp planning, and path planning. The basis for each bin-picker is a robust pose estimation approach—as nothing can be done until a relative pose of an object with respect to the robot is known. The field of research for this part of the system can generally be described as scene analysis. One of the first approaches published in the context of scene analysis is of 1963 [73]. But although many publications are very old, some of them have impact on the developments presented in this Ph.D. thesis and have to be mentioned when giving an overview on the *History of Bin-Picking*. A complete review on the wide field of research concerning the single subtasks of bin-picking would go beyond the scope of this thesis. Therefore, the referenced publications in the following sections shall be seen as a representative selection of the available literature. The author mainly focuses on publications which deal with the object localization problem as this is the most important part of a bin-picker. Other tasks, like collision avoidance, are important as well but can be simplified by properly designing the robot workspace. Experiments, described at the end of this thesis, show that very simple collision avoidance and path planning is sufficient for a robust and fast industrial bin-picker. Nevertheless, publications on these tasks will be part of the historical overview but not be treated in detail.

2.1 The Early Years: Basic Developments

As already mentioned above, many of the works that will appear in this summary are not dedicated bin-picking publications. They are rather dealing with one of the subtasks that have to be implemented for a complete picking system. In the beginnings of machine vision, robots were not as present as today, which is the reason for many publications to only mention the possible use of the approaches for robotics. Therefore, they offer developments on single topics that can be applied to several

© Springer International Publishing Switzerland 2016
D. Buchholz, *Bin-Picking*, Studies in Systems, Decision and Control 44,
DOI 10.1007/978-3-319-26500-1_2

other problems as well. Nevertheless, these works show basic and early approaches that have impact on modern developments, like those presented in this thesis.

Approaches of "blind" bin-picking in which robots, equipped with vacuum or magnetic grippers, mechanically scan their environment or the bin content until an object is grasped are not considered in the upcoming section, as these approaches are very slow and not robust. Automatic picking of objects stored in blisters will also be not part of the overview, as these situations can be handled by offline teaching of the robots and can therefore be solved without machine intelligence.

All other scenarios have to be handled by analyzing a 3D scene and extracting object poses from the available scene data. The first approaches used simple 2D images, as 3D scanners where not very common at the time of their publication. As already stated, the first work that shall be mentioned here is the Ph.D. thesis of L.G. Roberts from the year 1963 [73]. In this work former publications are cited, but the results of Roberts were very impressive and afterall, one publication has to be the first here. This thesis does not compute poses or even measure 3D data, but it is one of the first works estimating 3D shapes using single camera images. In detail, the goal is to detect polyhedral objects in a camera image and to re-render the scene from different viewpoints. Using the machine described earlier, Roberts computed a differential picture of the input to detect edges in the scene. With these edges, polygons were detected and analyzed to generate 3D polyhedrons. The generated 3D scene could then be viewed from different directions. One of the first works on 3D object detection was done in 1971 by Shirai and Suwa [81]. They developed a range finder that used a projector that was equipped with a vertical slit to project sheets of light onto the scene. With the known pose of this projector with respect to a camera, 3D point coordinates could be computed geometrically along the visible line. This development was an enhancement of a similar approach of Forsen in 1968 [32] which only projected light points and thus needed much more time for the depth image acquisition. These range finders were the only alternative to stereo vision systems in those days. With the acquired 256×256 depth image consisting of lines in 3D, Shirai and Suwa estimated those lines that were near edges of polygons and used these edge lines to estimate single planes. With these planes, polyhedrons could be located.

In 1975 Tsuji and Nakamura published a work that describes a vision system for an industrial application [88]. The goal was to classify and locate non occluded "complicated-shaped objects" in a pile of other objects in order to leave the "blocks world". Additionally, either the top or the bottom of the part had to be visible, further restricting its pose. To detect the objects, simple features like ellipses and border arcs are detected and compared to a model database. All processing is prone to errors due to lighting, occlusions and noise and does not consider depth information and thus is a challenge for further research.

As depth sensors were uncommon in the early years of pose estimation research, bin-picking as it is understood today was not easily manageable. But, when depth was unimportant, like dealing with objects lying on a known surface, complete object localization was possible. One of the approaches by Baird of 1977 is designed to handle objects lying on belt conveyors [13]. This system is not based on models and therefore computes the position and orientation of objects using their silhouettes and their centers of area as well as their axes of minimum moment of inertia. Even five years later the same assumptions were still made by Bolles and Cain [19]. They show a very similar method for localization of flat objects lying on a table using edge images. The objects needed to have features like holes and corners to be detectable. The same is true for the approach of Turney et al. [89].

In 1987 Perkins [70] made common assumptions: Known distance of the objects to the camera, rotations of the objects only in the plane, non-textured objects. He then used edge detection and built closed curves of the input image. By using a model data base of the objects, their poses could be computed via fitting of the centroids of corresponding curves. To solve the correspondence problem, the curves were described by up to 11 different features like total length, radius, magnitude of total angular change, etc. and compared pairwise. A very interesting publication in the context of this work is the paper of Ikeuchi, Horn and Nagata of 1983 [45]. The authors propose a method to estimate the object's orientation with respect to the camera using Photometric Stereo [99] and Extended Gaussian Images (EGIs) [39]. As these two techniques[1] play an important role for the developments described in this thesis it is very interesting that the basic ideas were developed over 30 years ago. Ikeuchi uses needle maps generated by the photometric method to build EGIs and uses these to estimate the orientation of the objects. The objects used in the experiments were tori and the overall method was limited to tori, as the special properties of them were used to solve several computations. The attitude estimation is based on a brute force comparison of the EGIs of a model torus and a measured needle map (normal map). To enhance the performance, the search area is reduced, exploiting the rotational symmetry of the tori. The position estimation is omitted by calculating a grasp point of the torus as a pixel within the 2D camera image. The grasp point is approached by a puma robot using the viewing ray of the camera, corresponding to the grasp point. By using a kind of photoelectric barrier in the gripper fingers, sending a signal, when an object is between the fingers, grasping of the object was enabled. All these limitations do not diminish the originality of the ideas of this work. Furthermore, this is one of the first complete bin-picking papers that included grasp planning and the actual use of a robot (PUMA). Unfortunately, until now only one further publication [40] (of the same authors) deals with the improvement of this system. Results of this work can be seen in Fig. 2.1. In the same year when Horn and Ikeuchi published their model-based approach, the group around Jean-Daniel Dessimoz published a different solution to the same problem [25] which is, along with the previous approach, very important for the developments in the present work. Dessimoz et al. proposed the use of matched filters to locate not complete models, but locations of object parts in an image that are accessible by a robotic gripper. The authors argued that not all degrees of freedom (DOF) need to be estimated for a successful robotic grasping. A very elegant aspect of their solution is that the localization procedure includes (very simple) grasp planning making an additional processing step obsolete. Based on an

---

[1] The Photometric Stereo method as well as Extended Gaussian Images will be topic of the following chapters.

(a)   **(b)**

(c)   **(d)**

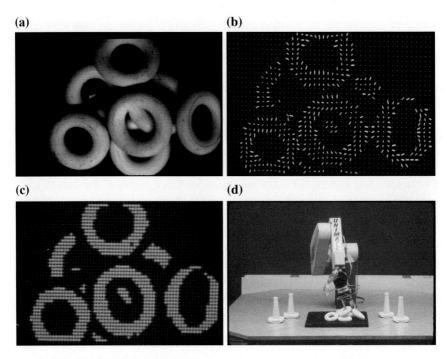

Fig. 2.1 Results of Ikeuchi's and Horn's work. **a** One of the three input gray level images. **b** Estimated needle map. **c** Segmentation result. **d** Robot grasping a located torus. All four images kindly provided by B.K.P. Horn

image of the actively lit bin, graspable regions are bright in the middle (an object is present that reflects the light) and dark aside (shadows due to the absence of objects that would collide with the gripper. To detect such regions, a simple filter kernel was used with which the image was correlated. The original kernel can be seen in Fig. 2.2. It was further rotated to not only estimate a fixed orientation for the gripper. Results of this work can be seen in Fig. 2.3. The interesting thing about this work is that only simple and standard image processing tools are applied to a gray level image to solve a very complex problem. The main disadvantage of the technique is that no depth information can be estimated and that the pose of the grasped object with respect to the gripper is unknown. This means that a defined placement, i.e. into a machine for further manufacturing, is not possible. As such a defined placement is essential for industrial bin-picking, this limitation is a big drawback. Nevertheless, a novel approach based on this publication will be presented in the following sections.

Bolles and Horaud described one of the first approaches to identify object locations in bins using 3D sensors and models of the objects, in 1986 [20]. They give a forecast of the importance and the spread of depth sensors in industrial applications in the future:

Fig. 2.2 The used matched
filter kernel of Dessimoz.
Image taken with permission
from [25]

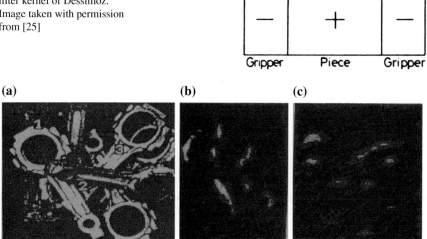

Fig. 2.3 Results of Dessimoz' work. **a** Input gray level image. **b** Estimated grasp points using
the filter kernel shown in Fig. 2.2. **c** Estimated grasp points using the filter kernel rotated by 90°.
In (**b**) and (**c**) the quality is denoted by the pixel brightness. The brighter the pixel, the better the
result. All images taken with permission from [25]

> Our approach is to use 3D models of the objects to find them in range data. Our rationale
> for this approach is that, first of all, range data simplify the locational analysis since the
> geometric information is encoded directly in the data. Secondly, it will soon be economical
> to use range sensors in industrial tasks. Finally, familiarity with the model of a part will
> add enough new constraints to make it practical to locate relatively complex parts jumbled
> together in a bin.[20]

As this is a very true forecast and the basis of this approach from 27 years ago is
exactly the same as used in this thesis, this paper is the last representative of "bin-
picking classics". The technique of Bolles uses a tree-search algorithm that extracts
single features, like edges, arcs, etc., out of a depth image. The feature type is selected
using the model of the part in the bin. Around each feature, other features are "grown"
that fit this specific feature. If some of the found features match the known model, a
hypothesis for an object pose is found and a depth image is rendered to evaluate that
hypothesis.

The last publication in this section is the Ph.D. thesis of Stahs of 1994 [86], which
was developed in the same institute as the present thesis and is used here as link to
modern approaches. Here, like in Bolles work, surface features are used to identify
and locate known objects. Simple features, like cylinders, planes, edges or holes
are combined to form more complex features with describing properties, like angles
between plane normals, distances between holes and edges, etc. These properties can
be stored in hash tables, using model data. By finding the hash keys of these complex
features (*local, minimal partial scene constellations*), pose estimation hypotheses

can be generated. This work builds the basis for the Ph.D. thesis of Winkelbach [96], which will be an important basis of the pose estimation technique, described in Chap. 3 and will be described there.

The literature presented in this section is far from being complete but gives a representative introduction to the roots of industrial vision and early bin-picking approaches. Some of the described publications even constitute a basis for approaches proposed in the following sections.

2.2 Modern Bin-Picking Approaches

During the years of research on object localization techniques, two main classes of approaches have been developed: Correspondence based and voting based techniques. The most prominent voting based technique is the well-known generalized Hough transform (HT) [14]. This approach is far from being modern but is used in this context to introduce the voting based methods.

Based on a model of the object to be located, the idea of the HT is that every sensed data point votes for all possible positions of the model's predefined reference point. The point in Hough space that gets the highest number of votes gives the coordinates of the model with respect to the sensor. The dimension of the Hough space is defined by the possible degrees of freedom of the object, e.g. six if the pose is not restricted at all. The resolution of the Hough spaces defines the possible accuracy of the localization. For each point of the scan, the complete set of model points has to be accumulated into the Hough space. The biggest conceptional problem of the Hough transform therefore is the high complexity with respect to the degrees of freedom of the possible objects pose and the amount of data.

A voting based technique that overcomes the complexity problems was presented by Mian et al. [64]. The authors modify the idea of geometric hashing proposed early in [57] by changing the hash tables to be filled with tensors instead of surface points. These tensors are computed using three-dimensional grids located at arbitrary positions on the meshes defined by a point tuple. The tensors are then constructed by estimating the surface area inside each of the bins of the grid. Each tensor, consisting of 10^3 bins is then stored in a 4D hash table which is spanned by the 3 grid coordinates and the angle between the normals of the point tuple. In this way problems of inhomogeneous point densities on meshes and scans can be reduced. Due to the complex computation of tensors, only 300 are used for the matching process. When many similar surface regions are present in the scan and/or model, this may lead to false matchings. Furthermore, this technique requires complex (pre-)processing steps, including mesh simplification and computation of intersections between the tensor grid and the mesh. This results in quite long computation times.

Correspondence based techniques use feature detectors and descriptors that are mainly known from feature tracking systems in 2D images like the well-known SIFT operator [60]. When these techniques are adapted to 3D data, they can be used for object localization. One example was developed by Zaharescu et al. who proposed a system containing a detector and descriptor for 3D mesh matching [101]. They interpret a mesh as 2D manifold in $\mathbb{R}^3$ to define gradients and convolutions on the mesh. Then, a function f can be defined on the manifold that maps the local mean curvature of the mesh or color (if available) to each vertex. Features are then detected by computing the difference of gaussians (DOG) of the function f and by extracting the maxima (the authors call this procedure "meshDOG"). To describe the features, a histogram of local gradients ("meshHOG") is computed around each feature. The meshHOG is built around a local coordinate system and consists of projections of gradients onto the three orthonormal planes, described by the coordinate axes, divided into 4 polar segments. In each segment, all gradients are stored in an eight-bin histogram. To locate an object, the procedure described above is applied to a scan and a model. By comparing the descriptors in a brute force manner, correspondences are found and pose hypotheses are generated. This approach and many other correspondence based approaches suffer from the issue that the features that they are based on have to be present on the models. If this is not the case, the approaches fail. If, for example, very simple objects, like a sphere or a cube, have to be located, no features may be found, or the extracted features may not be discriminative enough to be used for pose estimation. This is a general problem of correspondence based techniques.

If no model data is available, systems can not be based on model data, obviously. In this scenario, approaches to grasp unknown objects have to be developed. This thesis focuses on the bin-picking problem, where the task is to pick *and* place objects out of a bin. For this, model data has to be available. But, one of the approaches presented in Chap. 4 can also be used to grasp unknown objects. Therefore, a modern approach, using the same sensor data as in the experiments of this thesis by Fischinger and Vincze [29] shall be mentioned here, as well. Here, so-called height accumulated features ("HAF") are calculated. These are basically oriented and subsampled versions of the depth maps used. As for each possible gripper orientation, the depth map is converted into a point cloud, oriented and subsampled to an HAF, this method is very slow. An alternative method on this topic was published by Saxena et al. [78]. Here a complex machine learning approach based on synthetic 2D features is implemented. After the learning step, possible grasp points are located with a stereo camera setup and corresponding grasp points of both cameras are triangulated. In [15] camera images are used and analyzed to compute grasp regions. But as no depth information can be extracted of the 2D images, only simple table top scenes can be handled.

2.3 Yet Another Bin-Picking-Approach?

As described in the previous part of this work, the goal of a generic, robust and efficient bin-picking system is not met yet, even after decades of research in this field. There are, of course, several approaches that offer solutions to the bin-picking problem, or at least to single parts of the chain of specific needed operations. But, a complete solution describing a generic system cannot be found in literature, yet.

When the work on this dissertation started, industrial manufactures still had no possibility to buy a generic robotic bin-picker. An industrial applicable bin-picking solution was needed. With a wide overview of the literature, it became obvious that all of the known approaches have drawbacks or restrictions, or made assumptions that stood in contrast to the demanded generic system.

What is needed, is the development and implementation of a system that offers a solution in all single parts of the bin-picking process. The problems that needed to be solved were of industrial background and contained short cycle times, easy model exchange, robust pose estimation, secure collision avoidance and easy maintenance. Due to the fact that no system was available that met all these demands, vibratory feeders or human workers are still doing the bin-picking in modern industries.

2.3.1 Revisiting Robotic Bin-Picking—Problems to Be Solved

There are two main problems that have to be solved for bin-picking. The first and most important problem is *pose estimation*.[2] Three different poses are important in a bin-picking system:

- The *object pose* ^WP_O,
- the *gripper pose* ^WP_G and
- the *grasp pose* ^GP_O.

The object pose ^WP_O hereby describes the pose of an object with respect to some world coordinate system.[3] The gripper pose ^WP_G is the goal pose of the end effector w.r.t. the world coordinate system to grasp an object. The grasp pose ^GP_O is the pose of the object w.r.t. the gripper coordinate system, which is especially important for a defined placement of the object. These three poses build a transformation chain (see Fig. 2.4). Therefore, if two of them are estimated, the computation of the third is trivial.

Beyond pose estimation, a second important problem arises. This problem can be named as "*collision avoidance*". Each gripper pose has to be analyzed for possible collisions of the gripper with the environment. Only when the pose estimation and collision avoidance is solved, can robot movements be executed safely.

---

[2]Within this work, it is assumed that only one type of objects is in the bin. Otherwise, an object detection step has to be executed additionally.

[3]Within this work, it is assumed that the robot is calibrated relative to the world coordinate system. Furthermore, it is assumed that the vision sensor is calibrated to the same coordinate system.

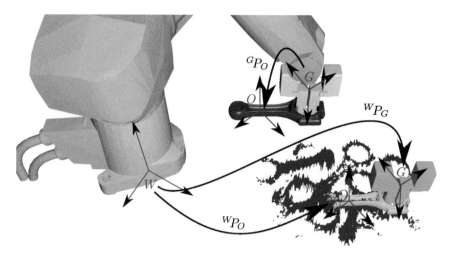

Fig. 2.4 Visualization of the three important poses in the workspace of a bin-picking robot. The object pose $^W\!P_O$ which defines the pose of the located object (*yellow*) in the world, the grasp pose $^G\!P_O$ which described the pose of the grasped object (*purple*) in the gripper and the gripper pose $^W\!P_G$ which is the pose of the gripper in the world (*green*) at which an object can successfully be grasped. It is assumed that the robot as well as the optical scanner are calibrated w.r.t. the world frame W. When the gripper grasps an object, the three transformations build a closed chain

2.3.2 Contributions and Organization of This Work

As already mentioned, the task of pose estimation is the basis of many applications in the field of robotic automation. Bin-picking is only one example of numerous tasks where the knowledge of accurate object poses is essential. These different tasks all may have different demands towards a pose estimation system. To make a contribution in this field that covers as many applications as possible, three different approaches towards solving the pose estimation problem are presented within this thesis. The main difference between the approaches is the sensor data the approaches are based on. To show the particular applicability of all approaches, each one is embedded into a bin-picking system and used to enable a robot to autonomously isolate objects scrambled in a box.

After the short overview on the tremendous amount of research on the topic of this thesis, given in Sect. 2.1, the end of this overview and short description of modern developments in Sect. 2.2 and a brief definition of the actual problem in Sect. 2.3.1, the rest of this work is organized as follows: The thesis is divided into three main parts. Each of these parts offers one way to solve the bin-picking problem, using a different pose estimation technique.

In Chap. 3, 3D point clouds are used as input data. How a known fragment matching approach can be applied to solve the localization problem of known models in 3D point clouds is described. This approach is then modified and enhanced to handle difficult real world scenarios in Sect. 3.1.2. The 3D pose estimation is augmented by a very efficient grasp planning mechanism in Sect. 3.2 and embedded into a robot work cell in Sect. 3.3.

The contributions in this part of the work are the adaptation of the Random Sample Matching algorithm to the pose estimation problem and its enhancement to work in difficult scenarios. A further contribution here is the semi automatic grasp pose estimation that offers highly flexible and efficient grasp planning and collision avoidance, respecting grasp pose restrictions possibly given by manufacturers.

In Chap. 4, a novel, complete bin-picking system based on 2D depth maps fused with inertia measurements is contributed. An extremely efficient, gripper pose estimation technique is described in Sect. 4.1 that computes collision free gripper poses nearly in real time. To enable it for defined placement of objects, this pose estimation is completed by a grasp pose estimation technique in Sect. 4.3 and embedded into a robot work cell in Sect. 4.4.

The contributions here are the proposed gripper pose estimation technique that generates poses in quasi zero time. Furthermore, the combination of this technique with a force/torque/acceleration based grasp pose estimation performed during the robot's movement results in the overall contribution of a nearly time optimal bin-picking approach.

The third contribution, Chap. 5 of this work, focuses on solving the pose estimation problem using normal maps. Normal maps were formerly only used to generate 3D data for further analysis. By introducing a technique to directly use normal maps for pose estimation, a whole group of sensors can newly be applied for automation tasks. Normal maps can be generated by very low-cost sensors using single camera shots which can result in very efficient systems.

The thesis is concluded in Chap. 6, where all three approaches are compared with each other and open problems are commented on.

Chapter 3
3D Point Cloud Based Pose Estimation

3D sensors are very popular in modern computer vision and are becoming cheaper. Therefore, many researchers focus on the analysis of 3D point clouds. Not only in the context of pose estimation but also in other fields of computer vision. For the pose estimation problem, a 3D point cloud seems to be a very good choice as sensor data. But, an efficient way to analyze a set of 3D points has to be found. This is the topic of this chapter.

At first, the impact of different sensor placements on the efficiency of bin-picking systems is discussed. Then, the pose estimation technique is explained in detail in Sect. 3.1. This approach is analyzed in the context of a real world scenario and a modification is presented in Sect. 3.1.2. To build a bin-picking system using the proposed pose estimation techniques, an efficient and semi-automatic grasp pose estimation is presented in Sect. 3.2. Both presented techniques are combined in an experimental setup in Sect. 3.3. A short discussion follows in Sect. 3.4.

Sensor Placement. Besides the choice of the sensor modality, its placement inside the work cell has to be chosen. This is not only important for the scanner used in the experiments, but also for every other optical sensor used for a robotic application. For this reason, these considerations are also true for Chaps. 4 and 5.

There are two main options to mount the sensor. It can either be mounted on the robot's end-effector or externally somewhere in the work cell. If an end-effector mount is chosen, no additional hardware is needed. The robot can be used to position the sensor w.r.t. the scene and a full scan can be acquired. But, if the sensor is mounted near the gripper of the robot, it has to be regarded during collision avoidance. The bigger the end-effector (including the sensor), the more likely collisions may occur between the robot and the scene. Therefore, it is likely that many objects cannot be grasped due to sensor collisions. Furthermore, when the robot is needed to generate sensor data, no pipelining of single subtasks is possible. For example, no new scan of the scene can be acquired during the object transfer movements of the robot.

Therefore, it is better to choose an external sensor mount. By this, the end-effector stays small and a new scan of the scene can be acquired during the robot's movements.

© Springer International Publishing Switzerland 2016
D. Buchholz, *Bin-Picking*, Studies in Systems, Decision and Control 44,
DOI 10.1007/978-3-319-26500-1_3

This leads to a more efficient performance of the system. Possible disadvantages of an external mount are the limitation to a single viewpoint and the need for additional hardware.

3.1   Generic Pose Estimation Using 3D Point Clouds

As a 3D sensor usually measures 3D coordinates at single points, a 3D point cloud has to be analyzed, when dealing with this kind of sensor. Inside the point cloud, the object has to be located, i.e. its six degrees of freedom have to be determined. In literature, different features are often used as search basis. These features can be planes [17], cylinders [77], edges of cylinders [67], box like geometric primitives [35] or others. The problem with using special features in a generic sense is, that there are always objects on which these features are not present and so new systems have to be designed.

Therefore, a solution has to be found that is not based on features at all, or only uses a kind of feature available on any possible object. In industry, the CAD models of the manufactured parts are usually available.[1] As the CAD models are present as, or can be easily converted to 3D point clouds, and the 3D scan is represented as point cloud, two identical data structures are available for the pose estimation.

The simplest features present in every model and every scan are single points lying on the surface of the object. There already exists a very efficient approach which compares surface features to solve the 3D puzzle problem [96, 97]. This approach will be adapted and used as generic pose estimation technique. The technique is described in the following section and is mainly extracted from [4].

3.1.1   3D Point Cloud Based Pose Estimation

When a tessellated point cloud (a mesh) is available, not only the coordinates of single surface points are known, but also a surface normal at each point can be calculated.

Models are usually available as meshes, the scans are not. But, as depth scanners always measure in a well structured way,[2] the meshing of the scanned point cloud is trivial (see Fig. 3.1).

When both data parts of the search process are present as oriented point clouds, the Random Sample Matching (RANSAM) algorithm can be applied to solve the pose estimation problem. It efficiently estimates the relative transformation between two 3D point clouds. No initial pose estimation is needed and no special features

---

[1] If this is not the case, the objects could be easily scanned.

[2] E.g. laser line scanners acquire depth information on equi-spaced points along a laser line which is translated across the scene, which leads to a regular grid of points. The same is true for depth cameras, which acquire depth information on a regular pixel grid.

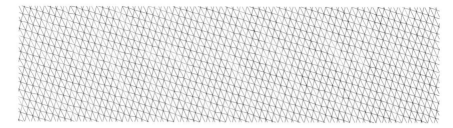

Fig. 3.1 Regular structure of a 3D mesh acquired by a laser line scanner. It is obvious that the tessellation can be performed in $O(1)$. The same structure exists when using ToF cameras because of their pixel grid

(except surface normals) of the objects to be located are used. For this reason, the RANSAM algorithm can be used for generic pose estimation, as it can easily be adapted to arbitrary object shapes by simply exchanging the CAD model.

The RANSAM algorithm is based on the well-known Random Sample Consensus approach (RANSAC) [30] exploiting the theory of the birthday attack [95]. In both meshes oriented point pairs $p_{u,v}$ (dipoles) are selected randomly. An oriented point in this case is defined as a 6D parameter vector $p_i := [v_i, n_i]$, i.e. a vertex with the according surface normal (see Fig. 3.2). For each dipole four translation and rotation invariant features can be calculated. These features are the distance d between the two points p_u and p_v, the angle δ between the normals n_u and n_v and the angles α and β between the connecting vector p_{uv} of p_u and p_v and the normals n_u and n_v respectively. The goal of the matching algorithm is to find two dipoles, one of the model and one of the scan, with four similar features. Two dipoles can uniquely be transformed onto each other. Therefore, by finding two equal ones, a pose hypothesis can be built. To efficiently search for the same dipoles, 4D relation tables are used. The axes of the tables are the four features d, α, β and δ of the dipoles. An individual relation table is assigned to both meshes. Then, dipoles are randomly chosen from both meshes in an alternating manner and, by their four features, stored inside the

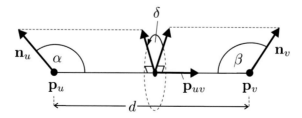

Fig. 3.2 Rotation and translation invariant features of a dipole. d is the distance between point u and point v, α is the angle between the normal n_u and vector p_{uv}, β is the angle between the normal n_v and vector p_{uv} and δ is the signed angle between n_u and n_v around p_{uv} (i.e. dihedral angle between the plane with normal $n_u \times p_{uv}$ and the plane with normal $n_v \times p_{uv}$). Graphic taken from [4]

according relation table. After storing a dipole, the same coordinate in the other relation table is checked for whether it already contains an entry with a dipole.

If the same key is found in both tables, a hypothesis for the relative transformation is found. Since the relation tables are filled continuously and only the table of the scan data has to be cleared after each manipulation of the scene, the pose estimation is very efficient.

3.1.1.1 Pose Hypotheses Generation

Let $\mathcal{A}$ be a mesh with according sets of vertices $\mathcal{V}_M = v_1, \ldots, v_k$ and normals $\mathcal{N}_M = n_1, \ldots, n_k$ of the model data and $\mathcal{B}$ be a mesh of the scan data, the goal is to find the relative transformation that 'correctly' fits the CAD model into the scan. One of the most important criteria for a good match is the amount of contact between the model and scan. An approach that considers a larger contact area of the scan promises more robust matching results than approaches that solely rely on local surface features.

Obviously, it is not efficient to exhaustively search through the 6d space of all relative locations. Therefore, only pose hypotheses with a certain surface contact between model and scan are considered. Such a hypothesis can be constructed by assuming a contact between some surface points. More precisely, four given oriented surface points $a, c \in \mathcal{A}$ and $b, d \in \mathcal{B}$ are sufficient, if a tangential contact between a and b as well as between c and d is assumed. This assumption constrains all degrees of freedom of the relative transformation. As illustrated in Fig. 3.3, the homogeneous 4×4 transformation matrix $^A T_B$ can be estimated by means of two predefined frames (one coordinate system for the CAD model an one for the scan):

$$^A T_B(a, b, c, d) \ = \ F(a, c)^{-1} \, F(b, d) \tag{3.1}$$

where the function $F(u, v)$ is a homogeneous 4×4 transformation matrix, representing a coordinate system located between the points u and v of a dipole

Fig. 3.3 The assumption of a tangential contact between two oriented point pairs can be used to define a relative transformation $^A T_B$. Graphic taken from [4]

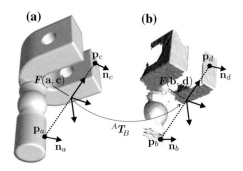

$$F(u, v) := \begin{bmatrix} \frac{p_{uv} \times n_{uv}}{\|p_{uv} \times n_{uv}\|} & p_{uv} & \frac{p_{uv} \times n_{uv} \times p_{uv}}{\|p_{uv} \times n_{uv} \times p_{uv}\|} & \frac{p_u + p_v}{2} \\ 0 & 0 & 0 & 1 \end{bmatrix} \quad (3.2)$$

with the difference vector p_{uv} and the combined normal vector n_{uv}

$$p_{uv} := \frac{p_v - p_u}{\|p_v - p_u\|}; \quad n_{uv} := n_u + n_v. \quad (3.3)$$

To avoid singular frames, it has to be ensured that the length of p_{uv} and n_{uv} is not zero. The calculated transformation $^A T_B$ aligns both dipoles. However, an exact tangential contact at two points is only possible if the relative distance of the points and the surface orientations at the contact points coincide. More precisely, it has to be ensured that the dipole (a, c) is geometrically congruent to the dipole (b, d). The relative transformation of one oriented point of a dipole to the other has four degrees of freedom. Therefore, at least four scalar quantities have to be compared to check for congruency. To verify this constraint, a 4D relation vector of a dipole is defined that consists of four values (one distance and three angles) that define the relative pose between two oriented points without ambiguity.

$$\text{rel}\,(u, v) := \begin{bmatrix} d_{uv} \\ \cos(\alpha_{uv}) \\ \cos(\beta_{uv}) \\ \delta_{uv} \end{bmatrix} := \begin{bmatrix} \|p_v - p_u\| \\ n_u \cdot p_{uv} \\ n_v \cdot p_{uv} \\ \text{atan2}\left(n_u \cdot (p_{uv} \times n_v), (n_u \times p_{uv}) \cdot (p_{uv} \times n_v)\right) \end{bmatrix} \quad (3.4)$$

These four values are illustrated in Fig. 3.2. Angle δ_{uv} denotes the dihedral angle between the plane with normal $n_u \times p_{uv}$ and the plane with normal $n_v \times p_{uv}$. Function $\text{atan2}(x, y)$ is similar to calculating the arctangent of y/x except that the signs of both arguments are used to determine the quadrant of the return value. The relation vector of Eq. 3.4 is invariant w.r.t. rotation and translation.

3.1.1.2 Rapid Generation of Likely Pose Hypotheses

In [96, 97] a highly efficient method for generating likely pose hypotheses by exploiting the theory of birthday attack [95]—an efficient cryptological strategy to generate two different documents with similar digital signatures (hash values)—has been proposed. In the following, this approach is summed up. Random dipoles of $\mathcal{A}$ and $\mathcal{B}$ are chosen, and alternately stored in relation tables (i.e. hash tables), using the four features of a dipole as table indices (see Fig. 3.4). Under the assumption that the invariants are unique, on average only $1.2 \cdot n$ pairs have to be processed until a collision occurs. This will provide a run-time complexity of $O(n)$ [97]. More precisely, the 4D relation tables (one per surface), and the four invariant features (Eq. 3.4) as table indices are used. This leads to the following search loop:

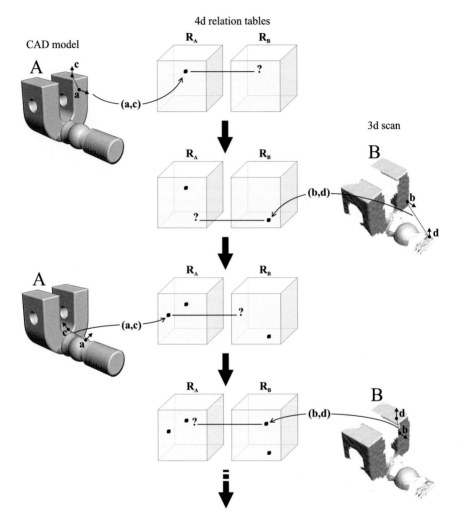

Fig. 3.4 Pose hypotheses generation using a 'birthday attack'-like approach: Random dipoles are inserted alternatingly into 4D relation tables. After a fiew processing cycles one can find dipoles with similar relations, which thus are geometrically congruent. Graphic taken from [4]

1. Randomly choose a dipole with $a, c \in \mathcal{A}$ and calculate rel(a, c).
2. Insert the point pair into the model's relation table: $R_A[\text{rel}(a, c)] = (a, c)$.
3. Read out same position of the scan's relation table: $(b, d) = R_B[\text{rel}(a, c)]$;
   if there is an entry $\Rightarrow$ new pose hypothesis (a, b, c, d).
4. Randomly choose a dipole with $b, d \in \mathcal{B}$ and calculate rel(b, d).
5. Insert the point pair into relation table: $R_B[\text{rel}(b, d)] = (b, d)$.
6. Read out same position of the model's relation table: $(a, c) = R_A[\text{rel}(b, d)]$;
   if there is an entry $\Rightarrow$ new pose hypothesis (a, b, c, d).

These steps will be repeated either until the hypothesis is good enough or all combinations are tested, or the processing time exceeds a predefined limit. The algorithm offers a run-time complexity of $O(n)$ for the first hypothesis, but since the relation tables get filled continuously, the complexity converges to $O(1)$ for further hypotheses.

3.1.1.3 Quality Estimation

After generating a pose hypothesis, i.e. a match between CAD model and scan, its matching quality has to be estimated. Therefore, the subset $\mathcal{C}$ of points on the surface of CAD model $\mathcal{A}$ which are in contact with the surface of scan $\mathcal{B}$ given a relative pose ${}^{A}T_{B}$ has to be computed. It is assumed that the surfaces are in contact at areas where the distances between surface points are smaller than a predefined ε_p:

$$\mathcal{C} := \left\{ a \in \mathcal{A} \mid \exists\, b \in \mathcal{B} \text{ with } \left\| p_a - {}^{A}T_B\; p_b \right\| < \varepsilon_p \right\}. \tag{3.5}$$

The tolerance value ε_p is necessary to handle noise and is therefore adapted to the surface accuracy. The quality is given by the ratio Ψ of contact points $|\mathcal{C}|$ to total number of surface points $|\mathcal{A}|$ of the surface of $\mathcal{A}$.

$$\Psi := \frac{|\mathcal{C}|}{|\mathcal{A}|} \tag{3.6}$$

The contact test is based on a *kd-tree* data structure (see [33]) and can therefore achieve a logarithmical time complexity for the closest point search. To increase the efficiency, Ψ can also be regarded as the probability that a random surface point $a \in \mathcal{A}$ is in contact with any surface point $b \in \mathcal{B}$. Thus, Ψ can be forecasted by an efficient Monte-Carlo strategy testing a sequence of m random surface points for contact. The advantage is that the quality estimation can be stopped early if the expected quality is considerably worse than the last best pose hypothesis. In this manner the quality estimation gets faster and faster, whenever the best hypothesis is improved. The disadvantage is that the optimal solution might get lost if the forecasted quality is very inaccurate. However, since many point pair combinations generate the same or nearly the same pose hypothesis, a loss of all good matches is very unlikely and thus can be neglected.

3.1.1.4 A Word to Practical Limitations—Field of View and "Scanability"

Regarding only the subtask of pose estimation and its solution presented above, in theory, a very versatile and generic solution is described. And, as the experiments in Sect. 3.3 will show, even under industrial conditions, the described approach performs very well. In the meanwhile, other researchers developed similar approaches based on the same technique [68, 69, 80]. But, there are cases with at least reduced robustness of the approach. In industry, robot (bin-picking) work cells can mostly be built using one depth sensor and a robot equipped with a gripper. This means that the scene is scanned from exactly one point of view. The robot itself could be equipped with a depth sensor, but this has many disadvantages, as mentioned earlier. For example, the end-effector becomes very big which results in a higher probability of collisions. And more important, robot movements are needed to generate the sensor data. So, the scanning procedure cannot be decoupled from the robot, making it impossible to use the time the robot moves the gripped object for new depth data acquisition. Time is one of the most expensive resources in industry. Thus, the eye-in-hand solution is inappropriate.

Practically, this means that the available depth data only covers a very limited part of the scene. By scanning from one direction, for example, faces that are parallel to the viewing direction of the scanner are not seen. If a triangulation based sensor is used, it can be even worse due to shadowing effects between e.g. laser line and camera. These shadowing effects can be due to bin borders or even shadows cast of objects onto themselves. For the most objects, all this has a negligible effect as shown in the experiments with e.g. piston rods.

With the goal of a real generic system, these issues have to be dealt with. An approach to overcome these problems is presented in the following section.

3.1.2 3D Edge Based Pose Estimation

In the previous section, it was mentioned that shadowing effects can be problematic for the described surface based pose estimation approach. Also, the former approach is sensitive to inhomogeneous point densities. This means that the surfaces of the scan and the model optimally have to have similar point densities over the complete surface. For example, when shiny metal parts, lying on a wooden plate, are scanned by a laser line scanner, the reflectivity of the metal results in many missing points whereas the wooden surface can be scanned properly (see Fig. 3.5). By locating a

Fig. 3.5 Scan of piston rods lying on a table (SICK LMS400). It is obvious that high matching values will be generated for poses that match the objects into the table surface

model with high point density in this scene, the model is likely to be found in the ground plane. And even worse, as the quality of the result is measured by the amount of vertices in contact, the false pose will be ranked better than any possible pose found on the real objects surface.

This problem can easily be avoided by cropping the search area. But this is only one aspect of the real problem. Other drawbacks cannot be avoided as easily. For example, if objects on a table are scanned from above and these objects consist of many planar faces with 90° angles, only edge vertices and planes are scanned and the same problem occurs (see Fig. 3.6). The problem is the same as before, namely the inhomogeneous point density.

Regarding the object shown in Fig. 3.6, another problem becomes obvious. Many possible dipoles found on this object will have the same relation vector. The relation tables will not get filled properly and many false pose hypotheses are the result. This problem is known in literature as self-similarity [7].

One possibility to overcome the described issues of inhomogeneous point densities and redundant relation vectors, is to extract regions of the scan which contain most of the information and to delete all other regions. Due to unknown object poses, the alternative to fill missing points is not possible in general. The interesting areas are the edges of the scan. As planar faces contain many similar dipoles, these areas do not fill the relation tables and thus will reduce the robustness of the quality estimation step. When wrong faces are matched onto each other, the matching will nevertheless get a high quality, as the edges only represent a small amount of points of the scan. So, the hypothesis generation as well as the quality estimation have to be modified for the described kind of scenario.

By extracting the edges of the scan data, only the most significant areas of the scan remain (of course, the model edges have to be extracted as well). Edge extraction in 3D is a complex problem (e.g. [51]). To efficiently solve this problem, the characteristic of the already mentioned single perspective scans that are present in

(a)   **(b)**

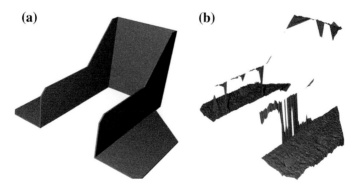

Fig. 3.6 Visualization of the occurring problem when sheet metal parts are scanned from a single point of view. **a** Model of a joist hanger. **b** Scan of the same object. It is obvious that the scan does not have a homogeneous point density (SICK IVP Ruler E1200)

most industrial scenes is employed. As the scene is scanned from one direction only, the scan data is well structured and can be represented by a 2D image. In this image, each pixel represents the distance of the scene to the sensor. These images are called *depth images* or *depth maps* and play an important role within this thesis. Using depth maps, the problem in 3D space can be reduced to a problem in 2D space. 2D edge extraction is a well-known problem in computer vision and many approaches exist. A standard gradient operator is completely sufficient here.

As the resulting 2D edges are edges in a depth image, also the 3D coordinates per pixel are known. By this, the point clouds can easily be reduced to edge points. As only vertices are available, the computation of surface normals is no longer possible. Even if the surface normals were estimated prior to the edge extraction, they would not be robust. Due to this problem, the algorithm described in the former Sect. 3.1.1 cannot be used for this scenario; the dipoles would only contain one feature (d) which is not descriptive enough for fast matching and even more important, would not contain enough information to calculate a pose hypothesis.

To overcome this issue a third point is added to the dipole used for hypotheses generation. On this *tripole* three features, d_{uv}, d_{uw}, d_{vw}, being the distances between single edge points, can be computed (see Fig. 3.7). The relation vector changes to

$$\text{rel}\,(\boldsymbol{u}, \boldsymbol{v}) := \begin{bmatrix} d_{uv} \\ d_{uw} \\ d_{vw} \end{bmatrix} := \begin{bmatrix} \|\boldsymbol{p}_v - \boldsymbol{p}_u\| \\ \|\boldsymbol{p}_w - \boldsymbol{p}_u\| \\ \|\boldsymbol{p}_w - \boldsymbol{p}_v\| \end{bmatrix}. \tag{3.7}$$

A similar adaptation has also been used in [46] where also no normals are available for the localization problem. The relative transformation of two equal triangles can then be calculated by estimating the relative transformation of the local coordinate systems defined on each of the triangles. The centroid of the triangle defines its origin. The x-axis points from the centroid to the middle of the shortest triangle side, the z-axis is defined by the normal of the triangle pointing towards the sensor and

Fig. 3.7 Rotational and translational invariant features of a tripole

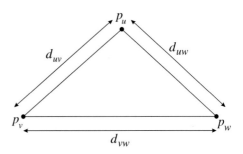

d_{uv}	distance between point $\boldsymbol{p}_u$ and point $\boldsymbol{p}_v$
d_{uw}	distance between point $\boldsymbol{p}_u$ and point $\boldsymbol{p}_w$
d_{vw}	distance between point $\boldsymbol{p}_v$ and point $\boldsymbol{p}_w$

the y-axis follows from these two. The poses of the two centroids $^{W}T_O$ and $^{W}T_S$ of the triangles of the object and the scan respectively then define the pose hypothesis: $^{S}T_O = \left(^{W}T_S\right)^{-1}{}^{W}T_O$.

3.2 Bin-Picking Application—Collision Avoidance and Grasp Planning

The last section described the task of pose estimation of known objects. So far, the robot would be able to approach the objects to be grasped. But, it has to be guaranteed that the execution of the pick movement is collision free. Therefore, when an object pose is present, a collision avoidance mechanism has to be applied to guarantee collision free movement of the robot. More specifically, a set of grasp poses has to be analyzed for collisions and the safest grasp pose has to be used to compute a gripper pose. One possible technique to cope with this problem is complete path planning of the robot as, for example, proposed in [91] where rapidly exploring random trees are used to find collision free paths through the workspace of the robot. Besides complex implementation, these approaches are often time consuming. Experiments using the system described in Sect. 3.3 and in a laboratory of Volkswagen Salzgitter have shown that a planned hardware setup is able to prevent the need for a complete collision avoidance on the robot's path. As in industry, robot cells are planned and built for special purposes, this assumption holds true for industrial applications. Besides that, the robot work cells are usually inside a fenced area which overcomes the problem of dynamic obstacles in the workspace. The experiments illustrated that it is enough to analyze the end effector pose at its computed grasp pose.

In industry, manufactured parts are often not allowed to be grasped at arbitrary locations. Furthermore, if objects are grasped at a not predefined pose, it is possible that it cannot be placed as desired and additional movements to regrasp the object have to be performed, as otherwise the gripper would collide with the machine or the conveyor where the object has to be placed. For these reasons, automatic grasp pose computation like proposed in [65, 74–76] is not part of this thesis. So, grasp poses have to be predefined and can therefore be limited to desired regions. To nevertheless allow for robust grasps, a semi automatic grasp pose evaluation is proposed in the following sections.

3.2.1 Efficient 3D Collision Avoidance

The problem of collision detection or better, collision avoidance can be described as: *"Given a grasp pose, estimate a measurement for the amount of collision between the scene and the end-effector at that pose."*

The scene (the scan mesh) and the model (the CAD mesh) are available as 3D point clouds, and the gripper can be represented as mesh in 3D, too. When an object is located, the gripper can be positioned relative to the object's coordinate system at its predefined pose. The straight forward solution for evaluating the pose and generating a collision measurement is to analyze the points of the scan mesh in relation to the gripper mesh. All points that lie inside the gripper indicate collisions of the gripper with the scene. The penetration depth, i.e. the distance between scan points and gripper surface can be used as collision measurement.

The algorithm can be described as follows. The surface normal for each gripper vertex is computed. For each vertex of the scan, the nearest vertex of the gripper is found. As a kd-tree is already available for the scan, this tree can also be used for a nearest neighbor search between scan vertices and gripper vertices. When the oriented gripper point p_g and its nearest neighbor of the scan p_s are found, the connecting vector p_{gs} of the two points is calculated.

$$p_{gs} = p_s - p_g \tag{3.8}$$

If the angle γ between p_{gs} and the normal n_g of vertex p_g is bigger than $\frac{\pi}{2}$,

$$\gamma = cos^{-1}(p_{gs} \cdot n_g) \geq \frac{\pi}{2} \tag{3.9}$$

the scan vertex lies inside the gripper. Then, the distance

$$d_c = |p_s - p_g| \tag{3.10}$$

can be used as collision measurement for this point. As the scan data is subject to noise and there may also be outliers, a simple boolean check for collisions is not optimal. Therefore, the sum of all single collision distances can be used as overall collision measurement E_G at that pose.

$$E_G = \sum_i d_{c,i} w_i \tag{3.11}$$

The factor w_i is a weighting factor which can be used to give different weights to different parts of the gripper. E.g., the "palm" can be weighted higher than the "fingers" as the fingers are always nearer to the objects and small finger collisions can be accepted due to a phased gripper geometry. If the penalty E_G is too high, the grasp pose is discarded. Furthermore, this penalty can be used to choose the best grasp pose out of all possible ones.

Besides the advantage of being very easy to implement, this technique suffers some drawbacks. For example, only single poses are evaluated, which means a high computational effort, when many grasp poses are tested. Furthermore, the gripper mesh has to have a very high resolution to enable correct measurements. This means computational overhead. But, it is possible to reduce the complexity and at the same time enhance the robustness and flexibility of the measurements. This enhancement is topic of the next section and implemented in a prototypical setup.

3.2.2 Depth Image Based Collision Measurement

The straight forward solution to the collision measurement mechanism is not very applicable due to its lack of flexibility. What is needed is a collision avoidance mechanism which, on the one hand, efficiently computes accurate collision measurements but, on the other hand, is not restricted to single poses. If only a small set of predefined grasp poses (defined relative to the object coordinate system) is used (like in the approach described above), it is likely that objects are not graspable that would be graspable if additional grasp poses would have been defined. A "semi-automatic" approach to grasp planning is introduced in the following paragraph, to overcome this issue.

3.2.2.1   Optimal Grasp Pose Estimation Using Key Grasp Frames

To enhance the performance of the technique described above, at first, the evaluation of single grasp poses is transferred from 3D point clouds to 2D images. The scan, as well as the gripper transformed to its grasp pose, are rendered as depth images. As the gripper is a closed mesh, it is divided into convex parts and each part is separated into its upper and lower surface. Now, the collision measurement is a simple pixel wise comparison of the depth images. Whenever a pixel of the upper face of an end-effector part is above the scan, but the same pixel of the lower face is located below, a collision in that pixel of the image is present. With the value of the differences of the pixels and the size of the pixels a collision volume can be calculated for this pixel. Whenever both surfaces are located below the scan, no clear evaluation can be done for that pixel because the gripper could be in occluded free space or in occluded collisions. The resulting unclassifiable volume is called *threat volume* (see Fig. 3.8). Collision and threat volume result in an overall collision volume

$$V_c = \sum_{\text{Part}} \sum_i s_x s_y d_i w_i. \tag{3.12}$$

Here, s_x and s_y are the pixel dimensions, d_i is the height difference of gripper surface and scan surface in the colliding pixel i and w_i is a weight which can be chosen to distinguish between collision and threat volume. Furthermore, w_i can give different weights to different parts of the gripper (Fig. 3.9).

At this point, it was shown how single grasp poses can be efficiently evaluated. To further enhance the performance, the approach is extended to work on predefined grasp regions. To achieve this, the **Key Grasp Frame** (KGF) is introduced by the author. A KGF consists of a base pose for the end-effector defined in the CAD

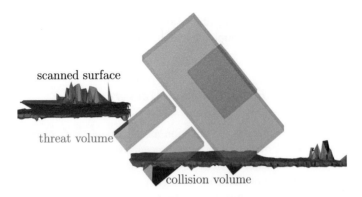

Fig. 3.8 Collision analysis of a gripper pose. The *yellow* volumes are unclassifiable threat volumes, as the *upper* and the *lower* surface of the gripper are *below* the scanned surface. Therefore, the areas can be occluded collision or occluded free space. The *red areas* are collision volumes, as the *upper part* of the gripper is *above* and the *lower part below* the scanned surface

(a)   **(b)**   **(c)**

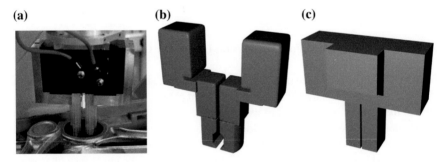

Fig. 3.9 The gripper used in the experiments. **a** Image of the parallel jaw gripper. **b** Complex CAD model used for point cloud based collision analysis. **c** Simplified model, sufficient and used for the KGF concept

model's coordinate system, a set of degrees of freedom (DOF) and an associated range for each DOF in which this base pose can be varied. This concept will still allow a wide variety of grasping positions (due to a quasi continuous variation of poses), while maintaining the demands of the industry to be in control of which region of the object classifies as a picking position. Furthermore, KGFs can easily be defined using the CAD model. An example for KGF definitions can be seen in Fig. 3.10. With the defined KGFs it is possible to evaluate each pose of the end-effector using the defined set of DOFs and their corresponding ranges. To minimize computational costs a new range image of both, the scan and the gripper is rendered using the gripper's coordinate system transformed to the base frame of each KGF. The z-axis of the gripper being the same as the approach vector serves as depth axis when translational DOFs are analyzed. For rotational DOFs, the specific rotation axis of the gripper is used as rendering direction, e.g. for grasps into circular holes, the approach axis is used. With this new coordinate system the scan as well as the gripper is rendered using orthographic projection (Fig. 3.11a, b). The rendered images can be

Fig. 3.10 *Key Grasp Frame* definition using a CAD model. KGF$_1$ contains a translational DOF, KGF$_2$ contains a rotational DOF. The gripper geometry for an inner grasp is different to that for an outer grasp

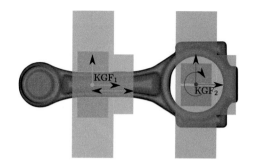

thought of as taken "looking through the gripper" using a virtual orthographic camera. The advantage is that the evaluation of the grasp poses and their ranges becomes a 2D problem. The area of the rendered scene is directly given by the type of the DOF (rotational or translational) and the range of the free parameter (Fig. 3.11c).

To analyze the translational DOFs, a simple correlation-like procedure of the scan and the gripper images is performed using the free DOF. The gripper images are shifted pixel-wise along the axis of the free DOF over the rendered scan image. Then, at each step, the rendered images are analyzed pixel wise for collisions as explained above.

The collision volume is stored for each step resulting in a 1D collision function dependent of the variable parameter. Using this function, all parameter values below a predefined collision threshold t_c are located and used to build a distance map for all possible collisions (Fig. 3.11e). If the gripper can be modeled as cuboids (Fig. 3.9), the described procedure can efficiently be implemented using integral images introduced in [92]. If not, efficient FFT based correlation can be applied to reduce computation times.

In case of a rotational DOF a preprocessing step is needed. To convert the rotational problem into a translational one, the rendered images are transformed into polar coordinates, which transforms the rotation into a shift along the orientation axis (Fig. 3.11d). The subsequent steps are then the same as before.

After the parameters for all defined KGFs are computed, the one with the highest distance to all collisions is determined using the calculated distance map (Fig. 3.11f). The collision volume resulting from this calculation is not as accurate as possible. Due to the (possibly) tilted point of view of the gripper relative to the scan direction, threat and collision volumes may have changed due to varying shadowing effects. Therefore, a second collision volume estimation has to be done using the original depth data for only the optimal frame computed in the previous step and calculating collision and threat volumes for the fixed final pose. Only if this final collision measurement is also below the threshold, the gripper pose will be selected for grasping. Otherwise, the next best solution of the prior step has to be used.

To visualize the concept, an example of the algorithm estimating an optimal pose using a rotational DOF KGF is shown in Fig. 3.11.

(a)   (b)   (c)

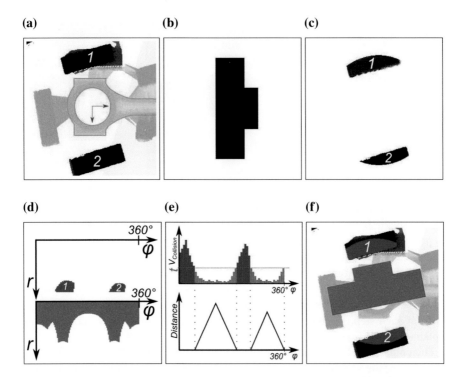

(d)   (e)   (f)

Fig. 3.11 Example for the optimal grasp pose estimation algorithm only using the palm of the gripper. The KGF is centered in the hole of the object (*inner grasp*) with a free rotational DOF around the approach axis of the gripper. **a** Depth image of the scan rendered in gripper coordinates, centered and oriented using the base frame of the KGF (*blue*) at a located object (*orange*) surrounded by obstacles (1 and 2) which would lead to collisions in certain gripper orientations. Dark values are near, bright values are far. **b** Upper depth image of the gripper palm model. **c** Section of the depth image, reduced by all pixels that do not collide with the gripper due to their height or distance to the KGF base frame. **d** Polar coordinate representation of the gripper palm (*green*) and the scan (*red*) respectively. This data is used to solve the best pose problem. **e** Collision function dependent on the rotational DOF (*above*) and the distance function to the nearest occurring collisions (*below*) with defined collision threshold t. **f** Superposed result of the optimal pose. Graphic taken from [1]

3.3 Experimental 3D Point Cloud Based Pose Estimation

In the previous section, a pose estimation technique was presented, based on 3D point clouds. To evaluate this technique as basis for an automated robotic task, a bin-picking station was prepared, consisting of an industrial robot and a 3D laser scanner.

In the experiments in this chapter as well as in the following chapters, all vision analysis was computed on a 3.6 GHz Windows PC with 8 GB RAM.

3.3.1 Simulation

To give a detailed overview of possible accuracies of the approach, the pose estimation technique was applied to a set of benchmark objects. Besides a set of industrial metal parts, the data set of [63, 64] was used. All models in the data set are scaled so that their longest edge equals 100 mm which fits the average size of the metal parts used in the real world scenario (see Figs. 3.12 and 3.13).

At first, the localization of the RANSAM algorithm was applied to full isolated models with different noise levels of $\sigma = 0, 7$, and 14 mm. The experimental results can be seen in Fig. 3.14 and show that the accuracy of the localization approach is very high, even in the presence of noise. It can also be seen that the computation time is dependent on the amount of vertices present in the two point clouds (scan and model) as well as the amount of noise. In a second series of simulated experiments, the RANSAM algorithm was analyzed for its robustness against noise and occlusions. A set of scenes of different objects scanned, using different viewpoints was used (see Fig. 3.15). In this set of images, only a small part of maximum 50 % of the objects was visible. As well as in the first set of scenes, the same amount of noise

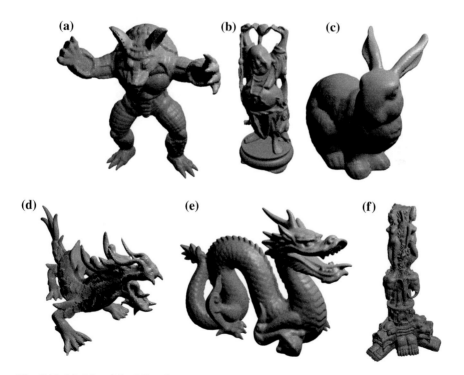

Fig. 3.12 Models of the Mian data set [63, 64]. **a** "Armadillo" consisting of 34834 vertices. **b** "Buddha" consisting of 32316 vertices. **c** "Bunny" consisting of 31064 vertices. **d** "Chinese Dragon" consisting of 36143 vertices. **e** "Dragon" consisting of 100207 vertices. **f** "Statuette" consisting of 40214 vertices

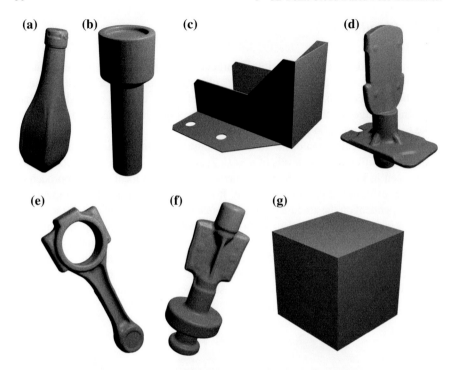

Fig. 3.13 Examples of models of the data set of industrial parts. **a** "Bottle" consisting of 143077 vertices. **b** "Cylinder" consisting of 414722 vertices. **c** "Joist Hanger" consisting of 90 vertices. **d** "Metal Part 1" consisting of 97636 vertices. **e** "Piston Rod" consisting of 58544 vertices. **f** "Balance Shaft" consisting of 81120 vertices. **g** "Cube" consisting of 728 vertices

was applied to these scenes. In total 5 viewpoints in 5 different scenes were tested. Each scene contained 3 different objects which means additional clutter. Each object was located 20 times. The results can be found in Fig. 3.16. Although, the scenes are very demanding (like locating the dragon in the scene shown in Fig. 3.15a) a valid object pose was computed in 99.4 % of the cases.

3.3.2   Real World Scenario

To test the applicability of the described localization approach a bin-picking station was built up. Within this station, the described localization approach, in combination with the KGF grasp planning method, was employed.

The bin-picking station consisted of a Stäubli RX60 industrial manipulator with an open control architecture [55]. The manipulator was equipped with a standard pneumatic parallel jaw gripper, as these grippers are rugged and therefore common in industry. Furthermore, a pneumatic overload protection device was placed between

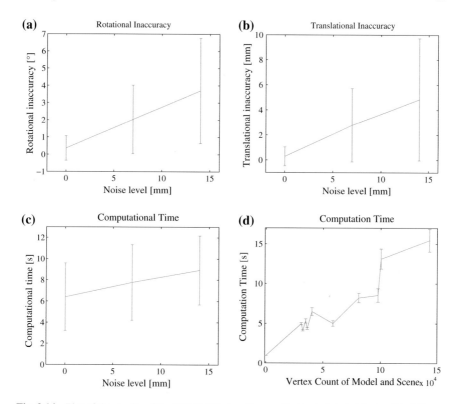

Fig. 3.14 Plot of the results of the RANSAM algorithm applied to isolated objects with different noise levels. The graphs show the mean and the standard deviation of the results. The applied noise is Gaussian noise with $\sigma = 7$ mm and $\sigma = 14$ mm. Besides a small decrease in accuracy, a slight increase in computation time is noticeable. **a** Rotational errors. **b** Translational errors. **c** Computational times dependent of the noise level. **d** Computational times dependent of the vertex count

the gripper and the wrist of the robot. For all experiments, the standard robot control delivered from the manufacturer would also have been sufficient. The robot itself was mounted in the work cell and its position was calibrated to a world coordinate system. Additionally, the base of the robot was placed in the same height as the upper border of the bin. This has the advantage that the robot arm does not collide with the bin borders which simplifies the collision avoidance and so allows shorter overall cycle times.

The vision hardware setup was built, taking the following aspects into account: The workspace contained a portal with a mounted linear axis. Different vision sensors were mounted on the sledge of the axis (see Fig. 3.17). This enables for easy comparison of the approaches described in the next chapters of this thesis. The height of the portal was high enough to avoid collisions between the robot and the portal as well as the sensors. Furthermore, the linear axis was placed in a way so that the viewing direction of the sensors was approximately the same as the approach direc-

(a)   **(b)**

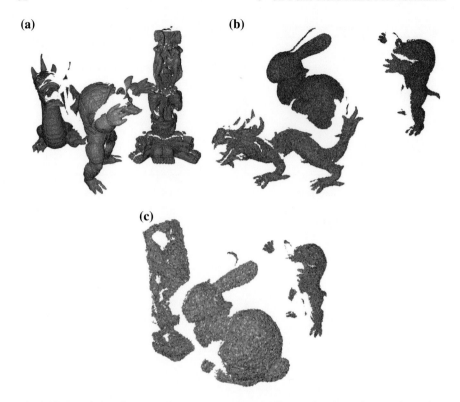

(c)

Fig. 3.15 Set of virtually scanned test scenes. **a** Scene without noise. It can be seen that only a very small portion of the dragon model is visible. **b** Scene with added noise of $\sigma = 7$ mm. **c** Scene with added noise of $\sigma = 14$ mm

tion of the robot when it was grasping parts and parallel to the bin borders. In this way as little problems due to shadowing effects as possible were present. Mounting all sensors on the sledge has the advantage that no sensor is ever in the field of view of another and all experiments can be performed using the same hardware setup.

The overall hardware setup was designed to be similar to possible industrial work cells.

For the 3D point cloud based pose estimation, which is subject of this chapter, two different sensors were used and analyzed for their applicability. These sensors were a *SICK LMS400* laser line scanner (data sheet available at [82]), which measures distances along a projected laser line using time of flight measurements and a *SICK IVP Ruler E1200* (data sheet available at [84]) which is a triangulation based laser line scanner.

Each of the two scanners has to be moved across the workspace to acquire a complete scan, as only the depth along one laser line is acquired at one location.

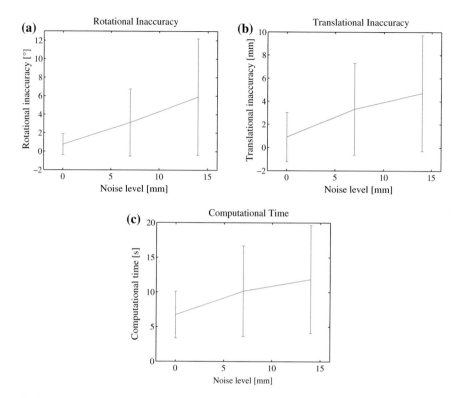

Fig. 3.16 Plot of the results of the RANSAM algorithm applied to the random viewpoint scenes. The graphs show the mean and the standard deviation of the results. The applied noise is Gaussian noise with $\sigma = 7$ mm and $\sigma = 14$ mm. Besides a small decrease in accuracy, a slight increase in computation time is noticeable. The experiments include very difficult scenes like shown in Fig. 3.15a, in which only small parts of the objects are captured. **a** Rotational errors. **b** Translational errors. **c** Computational times dependent of the noise level

   Both scanners have a wide scanning angle which, in combination with the movement of the linear axis results in a large scanned area. As the bin only covers a small part of this area, and the position of the bin is known, the area of interest (the content of the bin) is cropped, prior to vision analysis. This reduces the computation times because the amount of vertices is significantly smaller.

   As the bin is cropped out of the scan, a model of it is used for collision avoidance. In this way, it is assured that the bin borders, which are not scanned as they are vertical, are included in the collision avoidance.

   The two sensors mainly differ in the accuracy of the generated point clouds. The second important difference is that the *LMS400*, as it is not triangulation based and therefore does not need a big baseline, produces significantly less shadows at the cost of a lower accuracy. The data sheet gives a systematic error of ± 4 mm and a statistical error of ± 10 mm, which is comparable to the highest simulated amount of noise. Whereas the *Ruler* is specified with a typical height resolution of 0.4 mm,

Fig. 3.17 Three vision sensors mounted on the linear axis in the robot work cell. The moving Sick IVP Ruler providing 3D point clouds, a Microsoft Kinect providing depth maps, and an rgb-camera which, in combination with three light sources, provides normal maps

which is comparable to the noise free simulation. A visual comparison of the details captured by each of the scanners can be seen in Fig. 3.18. The *LMS* has an operating range of 0.7 . . . 3 m, the *Ruler* has an operating range of 0.28 . . . 1.28 m, i.e. both are suited for the experimental setup as well as for a possible industrial setup.

The first set of experiments was performed using the *LMS* with its high amount of noise. The second set of experiments was performed with the *Ruler*, which is more accurate. As objects served a set of industrial metal parts, which were scrambled unmixed in the bin. The task was to locate an object, to find a secure grasp pose and to place the object on a seating.As can be seen in the simulated experiments above, the amount of noise effects the accuracy and the computation time of the localization approach. The performance of the RANSAM algorithm is still very good and even with the LMS400, the experiments showed good results.

LMS400 Experiment Series. To test the applicability, the piston rod of the industrial parts data set was used. At first, the work cell was equipped with the LMS400. As no ground truth data is available, when the objects are scrambled in a bin, only the success rate of the pick and place cycles can be measured. Each localization took 4 s. The overall pick and place success rate was 95 % at 100 trials. The applied collision avoidance was only checking for collisions at predefined, fixed grasp poses defined relative to the models' coordinate systems. All unsuccessful grasp attempts did not result in collisions due to the collision avoidance mechanism but resulted in time overhead due to empty placement movements of the robot. The limiting factors in

(a) **(b)**

(c) **(d)**

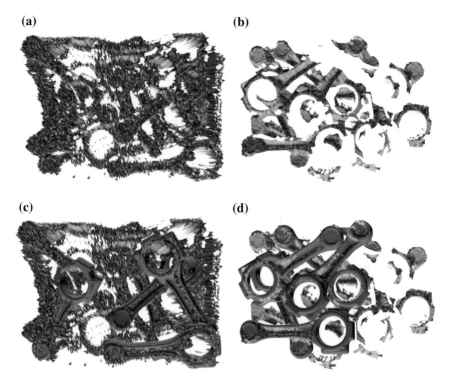

Fig. 3.18 Visual comparison of the scan quality of the SICK LMS and Sick IVP Ruler. **a** The scan of the LMS has less occlusions but a much higher amount of noise. **b** The Ruler has much less noise but a significant amount of holes caused by shadows between laser line and camera. **c** 5 located piston rods in the LMS scan, showing the robustness of the approach against noise. **d** 5 located piston rods in the Ruler scan, showing the robustness of the approach against occlusions

this setup were the limited accuracy of the 3D scanner and the collision avoidance mechanism. Therefore, after a short set of experiments these parts were replaced and a new set of experiments was performed.

Ruler E1200 Experiment Series. As the scanner accuracy effects both, the computation time and the accuracy of the localization, the more accurate Ruler 3D scanner was used in a second set of experiments, using the same objects. Additionally, the semi automatic grasp planner presented in Sect. 3.2.2 was included into the system. With these two changes, the grasp success rate of the system enhanced to 100 % at 100 reported trials.[3]

A success in this context means a collision free pick and place cycle and includes cycles where more than one object had to be located until a safe gripper pose was determined, which was the case in 14.1 % of the cycles. When a safe gripper pose

[3]The system runs as demonstrator at the Institut für Robotik und Prozessinformatik since the end of 2012. Furthermore, the system was transferred into a prototypical work cell at Volkswagen Salzgitter where it also showed promising results.

was found during first localization attempt, the overall computation time was 4.5 s with a standard deviation of 0.3 s. Including the cycles where more objects had to be localized, the average computation time was 5.5 s with a standard deviation of 2.6 s. The grasp pose estimation took about 0.1 s for a translational DOF and 0.3 s for a rotational DOF. This resulted in pick and place cycles in under 12 s. To achieve this cycle time, the scan procedure and the robot movements were pipelined. This means that the scan procedure started as soon as the robot left the space above the bin with a grasped object.

To evaluate the 3D edge based pose estimation, joist hangers were used as test objects and located in the bin. With the RANSAM algorithm, the localization success rate was only 38 % at 50 trials. Using the edge based RANSAM, 92 % of the localization results were correct. A comparison of the results in a worst case scene, with joist hangers placed directly on the ground plane of the bin, can be seen in Fig. 3.19.

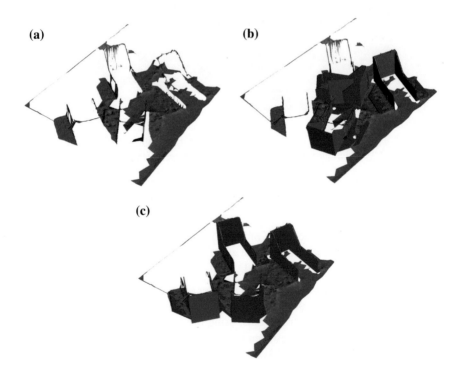

Fig. 3.19 Effect of the edge based localization attempt when locating critical planar objects. **a** Scan of a scene containing four joist hangers. The point distribution is not uniform, the vertical planes of the items cannot be scanned. **b** Matching attempt, using the face based RANSAM algorithm. Three of four localizations fail. **c** Matching result using efficiently extracted edges and the edge based RANSAM. All localizations are successful

3.4 Discussion

As the experiments in Sect. 3.3 show, the approach described above with its modifications is a very powerful way of solving the bin-picking problem for arbitrary objects. It was developed and built to meet all demands and restrictions of industrial applications.

Nevertheless, there is one main problem with this approach. The most expensive resource in industry is time. And so, the cycle times of manufacturing lines have to be as short as possible. This means, that the pick-and-place cycles of bin-pickers have to be as short as possible, too. The described system uses 3D laser scans of the scene. These laser scans have to be quite accurate so that the model based pose estimation performs robustly. If the bins are large and industrial sensors, like the SICK IVP Ruler are used, the scan time is already quite long. The scanner has to be moved over the bin to acquire the whole scene. Furthermore, in contrast to the parallelized cycles described in the experiments, for safety reasons, often the robot cannot move until the scanner has left the workspace of the robot. So, already several seconds are used just to acquire sensor data. Then, the complete object pose has to be estimated before the first movement command can be sent to the robot. If faster sensors like ToF[4]-cameras are used, the data acquisition time can drastically be shortened, but the resolution and accuracy of the point cloud is no longer high enough for the pose estimation to perform robustly.

In summary: When time is not a big issue, the 3D point cloud based pose estimation technique is successfully applicable as solution for the bin-picking problem for arbitrary objects and for object localization in general. If time is an issue, the system may perform too slowly.

In such a case, an alternative approach has to be found which computes motion commands for the robot so that the next movement is available at every time in the cycle. This would lead to the shortest possible cycle time as the robot's movements are the only limiting factor which cannot be avoided in any way. The experiments of the collision avoidance mechanisms described above already showed that the reduction of point clouds to depth images is a very promising step toward generating efficient 3D algorithms. In this way, the 3D scanner's structure is considered and used for data simplification. If the complete computations could be based on 2D depth images only, a very efficient system could be the result. Approaches using exactly this are subject of the next section.

---

[4]Time of Flight.

Chapter 4
Depth Map Based Pose Estimation

As already mentioned in Sect. 3.1.1, (industrial) scenes are often scanned from one direction. Furthermore, scan data from e.g. a laser scanner or depth camera, are almost always well structured. This means that either the sensor of the depth camera defines a grid on which the data points lie, or the grid is spanned by the laser line and the linear axis, when using a laser line scanner. In Sect. 3.1.1 this led to the approach to generate depth images and to use these to efficiently estimate 3D edges to perform the matching procedure on distinct parts of the scan. For collision avoidance, the understanding of the 3D point clouds as 2D depth images resulted in a very efficient way of semi-automatic estimation of grasp poses.

Taking this one step further, the use of point clouds seems to be an unnecessary overhead for these types of scenarios. Or, in other words, building a point cloud using a 3D sensor scanning from one direction means to convert data ordered on a 2D grid into 3D data with no gain of information but an increase of complexity. Additionally, the acquisition of a depth map, using a 3D camera is by far less time consuming than using a laser line scanner. So, the data acquisition time in a bin-picking system could be reduced drastically by using a single shot depth camera as already mentioned. As time is the most expensive resource in industry, time saving at the cost of sensor data accuracy can be a good deal if it is possible to overcome the higher data noise by robustness of the applied algorithms and if the image analysis is fast enough to make full use of the short acquisition times.

When the data acquisition and analysis are very time efficient, the work flow of the complete system can further be improved to achieve short cycle times. The system of Chap. 3, like many other bin-picking approaches, at first estimates the 6D pose of an object, and then computes a gripper pose at which that object can be grasped. So, in every cycle, the complete object pose is computed before the robot starts the transfer motion to place the part at the desired position. But in fact, in many cases[1]

---

[1] Precise object poses are needed, in cases where objects are only allowed to be grasped at predefined regions.

© Springer International Publishing Switzerland 2016

D. Buchholz, *Bin-Picking*, Studies in Systems, Decision and Control 44,

DOI 10.1007/978-3-319-26500-1_4

the object pose is not needed at all to start grasping. It is rather interesting to find a gripper pose where the robot can grasp an object. Only when the robot has the object in its gripper and has completed the depart motion, the object pose inside the gripper (the grasp pose) is needed to place it in a defined way.

Therefore, a direct computation of essential information is preferable to an indirect computation to avoid time overhead. Again, in the context of the 3D point-cloud-based bin-picker, the gripper poses are indirectly computed using object poses and it is possible that located objects are not graspable. In such cases, some other object has to be located. If a gripper pose is computed directly, this case may not occur.

The following process schedule results from the presented considerations:

1. Acquire a depth map of the workspace.
2. Find a graspable region in the scan data.
3. Approach and grasp the related object at the computed gripper pose.
4. Execute depart motion.
5. Estimate the pose of the grasped object inside the gripper.
6. Place the object in a well-defined way.

The rest of this chapter describes these steps and is organized as follows. In Sect. 4.1 a very efficient way of gripper pose estimation is presented. Some concepts to enhance the proposed gripper pose estimation technique are presented in Sect. 4.2. Then, in Sect. 4.3, the system is completed by a grasp pose estimation technique which enables a defined placement of the grasped object. Section 4.5 concludes the chapter by discussing the described bin-picking technique.

4.1   Gripper Pose Estimation

The collision avoidance mechanism described in Sect. 3.2 already works on 2D images. There, the scan mesh as well as the gripper were rendered using the scanned point cloud and the approach direction of the gripper to analyze the collision volume and to reduce the dimension of the problem from 3D to 2D. It became obvious that when the grasp procedure was not performed exactly parallel to the scan direction, shadowing effects occur and may alter the quality of the collision measure. To avoid this issue, all grasping movements can be performed using the viewing direction of the optical scanner as approach direction for the gripper. Doing this, no shadowing effects may lead to undetectable collisions, or to use the KGF concept: no threat volume will occur. To be able to do this, the scanner has to be mounted in a suitable position relative to the robot.

Since the scene information is present as a gray level image, standard image analysis techniques can be applied to solve the pose estimation problem. As gray level images were the basis for the bin-picking approaches first addressed 30 years ago (see Sect. 2.2), it is still useful to look into these classic papers. In [18] an approach for gripper pose estimation using matched filters was presented in 1984. Surprisingly, there have not been further developments of the ideas presented in this work. By adaptation of the methods of this classic paper to depth images a very efficient gripper pose estimation technique can be obtained.

4.1.1 Fast Gripper Pose Hypotheses Generation

To revise, a gripper pose $^W\boldsymbol{P}_G$ is a pose of the gripper relative to the world coordinate system, at which an object can be grasped. Using depth images as model for the workspace, this means that a gripper pose is a feature in the depth image that shows a local decrease in depth, therefore a local increase in object height. In other words, a pattern must be found in the image which matches the gripper footprint in its appearance. The gripper used in the experiments is a standard parallel jaw gripper that was already used in former experiments (see Fig. 3.9). The footprint of the gripper fingers, which are the important part of it for grasping, can be approximated as two simple squares. As the gripper may be rotated around its approach vector and the algorithm shall be as fast as possible, a filter kernel is used that is rotationally symmetric and thus contains all possible gripper orientations (see Fig. 4.1). This simplification ensures a gripper pose estimation via a single convolution operation. Such a filter kernel can also be used for other gripper types. No matter if a multifingered hand, or an angular gripper is used, the kernel stays the same.

The correlation C of the depth image I_D and this kernel K'_G yields a set of gripper pose hypotheses as these are the local maxima of the correlation function. K'_G denotes a possibly scaled version of K_G. In the case of a perspective depth camera like used in the experiments (Microsoft Kinect), the size of the pixels is dependent on the distance of the surface to the camera. This distance is approximated by using the closest pixel in the depth image (see Sect. 4.4).

$$C(x, y) = \sum_{\xi} \sum_{\eta} I_D(\xi, \eta) \cdot K'_G(\xi + x, \eta + y) \tag{4.1}$$

The locations (x_m, y_m) of the maxima of $C(x, y)$ define two DoFs of the gripper pose in pixel coordinates. These coordinates then have to be transformed into the sensor coordinate system and evaluated and analyzed to find the optimal valid gripper

Fig. 4.1 Gripper kernel K_G. **a** Signs of weights. **b** Optimal graspable object superimposed. Graphic taken from [2]

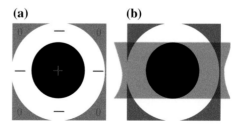

(a)   **(b)**

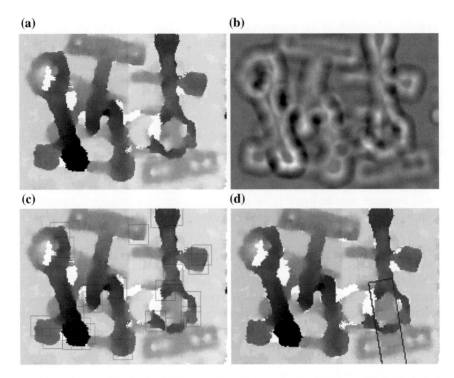

(c)   **(d)**

Fig. 4.2 Pick pose estimation by correlation. **a** Depth image of a pile of different objects. *Dark pixels* are closer to the camera. **b** Correlation result C. Regions that are high and have low surroundings produce the highest (*brightest*) results. **c** Generated hypotheses using local maxima of C. The best maximum (*green square*) is further analyzed in Fig. 4.3. **d** Valid pick pose shown as superposed gripper footprint. Graphic taken from [2]

pose. The third DoF can be found by the local depth $z_g = I_D(x_m, y_m)$ in the image (see Fig. 4.2).

4.1.2 Hypothesis Evaluation and Gripper Pose Estimation

Each maximum in the resulting correlation image represents a hypothesis for a gripper pose ${}^W\boldsymbol{P}_G$. But, only three parameters of this pose have been estimated. Due to the orientation invariance of the gripper kernel, a suitable orientation of the gripper is still unknown. Furthermore, it has to be evaluated, if and how far the pose can be approached without collisions.

The rotation angle around the approach vector can be computed using a local patch of the image around the maximum. A local threshold is applied to the patch, sized like the filter kernel, to generate a binary image in which the maximum values are set to 1 and the minimum values to 0. This separates the graspable object region from the "background". The binary image now has to be analyzed for optimal finger placement. Depending on the gripper type, different procedures can be applied. In the case of a parallel jaw gripper or angular gripper, the orientation can be estimated as follows.

The topological skeleton of the binary patch is computed. Using only the skeleton, a line can be fitted into the patch using RANSAC [23] or deterministic linear regression techniques. The normal of the estimated line describes the optimal orientation Φ of the gripper to grasp this region (see Fig. 4.3). If a 3-jaw gripper is used, a correlation could be used to find the best orientation which is similar to the procedure described in Sect. 3.2.2. For a multifingered hand, the free space around the grasp hypothesis could be analyzed for possible finger positions. The analysis for a vacuum gripper is trivial. It is suitable to find a planar region and no orientation has to be estimated. Now, only one additional parameter, the approach depth d_a, has to be found. This parameter contains information about how far the gripper may approach the gripper pose and can easily be estimated using the pixels at the computed jaw positions. In this context, the jaw position means the hull of the footprint of the fingers during their close or open motion. Subtracting z_g from the minimum depth of all pixels covered by the gripper footprint yields d_a. Figure 4.4 illustrates its computation.

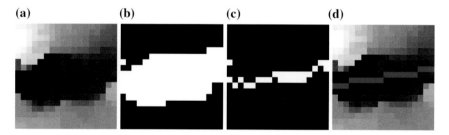

(a)   **(b)**   **(c)**   **(d)**

Fig. 4.3 Estimation of the gripper orientation Φ of the marked hypothesis in Fig. 4.2 for a parallel jaw gripper. **a** Section around pose hypothesis (x_m, y_m). **b** Binary image of the section. The object is separated from the background. **c** Topological skeleton of the section. **d** Estimated line using RANSAC. Graphic taken from [2]

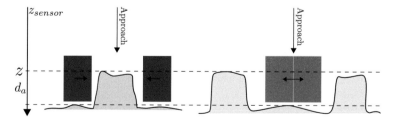

Fig. 4.4 Calculation of the approach distance d_a using outer grasps or inner grasps of a parallel jaw gripper. The same approach can be used for any type of gripper. Graphic taken from [2]

With the algorithm described above, many hypotheses are generated. All valid gripper pose hypotheses have to be ranked to find the best one to approach. One quality measure is the parameter d_a, which is already available, and assures that no collisions occur between the gripper jaws and the scene. As d_a becomes larger, the grasps at the objects get more robust and shall be prioritized. Additionally, it has to be checked that d_a exceeds a certain threshold ϵ_p and that the gripper palm does not collide with any obstacle. The threshold ϵ_p has to be set to fit the gripper properties. By analyzing the palm footprint, which, in the case of the gripper used in the experiments, is a simple rectangle in the depth map, in the same manner as d_a has been obtained, full collision avoidance can be guaranteed. A further quality measure is z_g which is the depth of the gripper in the bin. Objects that lie higher than others should be grasped first. Lower z_g therefore results in higher ratings of the gripper pose hypothesis. Other quality measures can be applied if needed.

The described steps to compute a valid gripper pose are summarized in Algorithm 1. Note that the output $^W\boldsymbol{P}_G$ does not contain the pitch and yaw angles since they are fixed by the sensor setup as described above.

Data: depth image I_D, gripper-defined kernel K_G
Result: Gripper Pose $^W\boldsymbol{P}_G = \{x_g, y_g, z_g, \Phi\}$, d_a
find global minimum M_d in I_D;
scale K_G according to M_d;
Correlation function $C = I_D \star \star K'_G$;
while *no valid pose found* **do**
    find global maximum M in C;
    $x_m \leftarrow M.x$;
    $y_m \leftarrow M.y$;
    $z_g \leftarrow I_D[x_m, y_m]$;
    estimate topological skeleton line l in patch around (x_m, y_m);
    $\Phi \leftarrow l.normal$;
    estimate approach depth d_a;
    if $d_a > \epsilon_p$ **then**
       if *gripper palm is collision-free* **then**
          valid pose found;
       else
          continue;
       end
    else
       continue;
    end
    delete local area in C around (x_m, y_m);
end
transform (x_m, y_m) from pixel values into world coordinates;
$x_g \leftarrow T(x_m)$;
$y_g \leftarrow T(y_m)$;

Algorithm 1: Gripper pose estimation.

4.2 Modifications and Enhancements

So far, not all possibilities of the approach are fully utilized. It is mentioned how grasping from outside (grasping by closing the gripper jaws) using four DOFs of the robot's end-effector is possible. Many enhancements to this approach are possible. If, for example, the objects to be grasped contain regions enabling a grasping from inside, the local minima of the correlation function can be used as inside gripper pose hypotheses. It may be necessary to adapt the filter kernel for a grasp from inside, depending on the gripper geometry. In this case, a second convolution would be necessary. In most cases, it is enough to decide for one pick style and use that to clear the bin.

Different gripper types were already mentioned above and are a further possible modification which can easily be integrated into the proposed system.

To enhance the robustness of the grasps, further analyses of the depth image are possible. If the objects handled are graspable at their very ends, situations may occur where two endpoints touch each other and two objects would be grasped accidentally. To overcome this issue, a connectivity analysis of the binarized local patch has to be done, to dismiss these regions. For inner grasps, a very similar procedure has to be performed, to avoid grasps between separated objects. For this, a complete contour can be detected in the region of the grasp hypothesis to assure validity of the gripper pose.

More complex modifications are described in the following sections.

4.2.1 Pitch and Yaw Angles of the Pick Pose

So far, the rotation estimation of the gripper was limited to the rotation around the approach axis, i.e. the roll angle Φ. Special types of objects may require a more accurate grasp pose. For example, if the objects contain many cylindrical regions, Φ is enough to estimate for a secure grasp, but an unaligned grasp of a tilted planar object region might result in damages at the gripper or object (see Figs. 4.5 and 4.6). In this case, pitch and/or yaw angles have to be computed additionally to the roll angle of the gripper. This can be achieved by local analysis of the depth patch in the grasping region. By approximating a plane in this region, i.e. by principal component analysis, a normal can be computed describing the average orientation of the object. The open vector o of the gripper must then be aligned with the plane normal n.

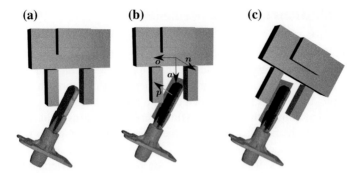

Fig. 4.5 **a** Example of an object that should not be grasped with fixed grasping orientation. The object or the gripper would be damaged during the grasp procedure. **b** Estimation of a plane in the grasp region. **c** The open vector of the gripper coordinate system has to be aligned with the normal of the plane to ensure safe grasping

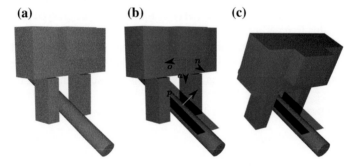

Fig. 4.6 **a** Example of an object that could be best grasped with a variable pitch angle of the gripper. The grasp on this object would be more robust. **b** Estimation of a plane in the grasp region. The plane estimation works, because only the upper side of the object is scanned due to the single point of view of the sensor. **c** The approach vector of the gripper coordinate system has to be aligned with the normal of the plane to ensure best grasping

4.3 Bin-Picking Application—Grasp Pose Estimation

So far, objects may be grasped from a pile or a bin and may be transferred to a target location coarsely defined, e.g., using the joint angles of the robot. For industrial applications or service robotics, it is often essential to place the grasped objects at a predefined *position* in a desired *orientation*. In these cases, model data of the objects are required.

Employing the gripper pose estimation approach presented in Sect. 4.1 has the drawback that the pose of the grasped object w.r.t. the gripper (denoted as grasp pose) is unknown. Regarded from a different point of view, it has the essential advantage that only the parameters *needed* for grasping are determined *prior to* grasping. Thus, the grasp motion could be initiated earlier; idle times of the robot could be reduced and therefore pick and place cycle times could be minimized.

Nevertheless, to be able to place the grasped object at a desired pose, the grasp pose has to be determined *before* the object is *placed*. In this section, approaches are presented that estimate the grasp pose of the object. This happens either using the sensor data used for gripper pose estimation, or *during* the transfer motion *after* the object has been grasped, using force/torque and acceleration sensors. In fact, by estimating the coordinates of the center of mass of the object and its principal axes of inertia, a finite set[2] of pose hypotheses can be obtained.

The next section describes a technique for grasp pose estimation based on visual data. Then, a procedure to estimate the inertial parameters of an object which has been grasped will be presented. The derivation of robust pose hypotheses from inertial parameters is addressed and strategies to deal with pose ambiguities are proposed.

4.3.1 *Vision Based Grasp Pose Estimation*

After the gripper pose is estimated, the robot can start the approach and grasp movements. The time needed to perform these movements can be used to further analyze the depth image. Until now, the estimation of the object pose $^{W}\boldsymbol{P}_O$ was avoided as it was not needed to grasp the object. Now, as a gripper pose $^{W}\boldsymbol{P}_G$ is already available and the next information needed is the grasp pose $^{G}\boldsymbol{P}_O$, the object pose is of interest. By estimating the object pose, using the known gripper pose, the grasp pose can easily be computed by closing the transformation chain

$$^{G}\boldsymbol{P}_O = \left(^{W}\boldsymbol{P}_G{}^{-1}\right) {}^{W}\boldsymbol{P}_O. \tag{4.2}$$

The direct estimation of $^{W}\boldsymbol{P}_O$ is topic of Chap. 3 and can be used here as well. A local patch of the depth image can easily be transformed into a 3D point cloud. By using a region growing technique with a pixel of the depth image between the gripper fingers as a seed pixel, a segmentation can be used to build this patch that then nearly only consists of points of the grasped object. A RANSAM match can then be computed very easily. The time that passes until a possible target pose for the object is reached can be used for this computation, which means that the pose estimation can be performed simultaneously to the robot's transfer movements.

A problem that may occur is that the object may move during the closure of the gripper fingers. This mainly happens, when pneumatic grippers are used without tactile sensors built in the fingers. If this happens, the estimated gripper pose is only a coarse estimate of the real pose. To overcome this problem, further sensors can be included into the system.

---

[2]Not considering ambiguities arising from symmetry.

4.3.2 Force/Torque/Acceleration Based Grasp Pose Estimation

When an object is grasped by a manipulator and moved from one point to another, forces and torques are exerted on the wrist of the robot. These forces and torques are caused by the inertia of the object and the gripper. By measuring these forces and torques, information about the pose of the object inside the gripper can be acquired, if model data is available. To acquire this information, equations can be derived using the basic laws of dynamics; the complete set of inertial parameters can be estimated. These parameters are the product mc of the object's mass m and coordinates c of the center of mass (COM), and the elements of the inertia matrix I.

Based on these parameters, four features can be derived that are invariant to rotation and translation. These four features are the mass and the three principal moments of inertia; they can be estimated using sensor data acquired during the movements of the robot. A force/torque sensor as well as an acceleration sensor are mounted between the wrist of the robot and the gripper. These sensors deliver 6D force/torque values and 6D acceleration values. 3D angular velocity values are measured by an inertial measurement unit (IMU). Additionally joint angle data is acquired.

Now, based on the Newton Euler approach, the dependence of the external forces and torques and the motion of an object can be described by two vector equations that are linear with respect to the unknown parameters.

$$^S\!f = m^S\!a - m^S\!g + {}^S\!\alpha \times m^S\!c + {}^S\!\omega \times \left({}^S\!\omega \times m^S\!c\right) \tag{4.3}$$

$$^S\!\tau = {}^S\!I^S\!\alpha + {}^S\!\omega \times \left({}^S\!I^S\!\omega\right) + m^S\!c \times {}^S\!a - m^S\!c \times {}^S\!g \tag{4.4}$$

Here, the superscript ^{S} indicates that the sensor's coordinate system S is used as reference frame. $^S\!f$ and $^S\!\tau$ are the measured forces and torques. $^S\!a$ and $^S\!\alpha$ are the linear and angular acceleration vectors and $^S\!\omega$ and $^S\!g$ are the angular velocity vector and the gravity vector. $^S\!I$ is the inertia matrix with

$$I = \begin{pmatrix} I_{xx} & I_{xy} & I_{xz} \\ I_{xy} & I_{yy} & I_{yz} \\ I_{xz} & I_{yz} & I_{zz} \end{pmatrix}. \tag{4.5}$$

The elements of matrix I constitute the moments and products of inertia. The complete set of ten inertial parameters can be compiled into a vector

$$^S\!\varphi = \left[m, m^S\!c_x, m^S\!c_y, m^S\!c_z, {}^S\!I_{xx}, {}^S\!I_{xy}, {}^S\!I_{xz}, {}^S\!I_{yy}, {}^S\!I_{yz}, {}^S\!I_{zz}\right]^T. \tag{4.6}$$

With $^S\!\varphi$, Eqs. 4.3 and 4.4 can be written in matrix form[3]

---

[3] Note that the index ^{S} is omitted in Eqs. 4.9 and 4.10.

$$\begin{pmatrix} {}^S f \\ {}^S \tau \end{pmatrix} = V\left({}^S a, {}^S g, {}^S \omega, {}^S \alpha\right) {}^S \varphi \tag{4.7}$$

with

$$V = (V_{mc} \, V_I) \tag{4.8}$$

and

$$V_{mc} = \begin{pmatrix} a_x - g_x & -\omega_y^2 - \omega_z^2 & \omega_x\omega_y - \alpha_z & \omega_x\omega_z + \alpha_y \\ a_y - g_y & \omega_x\omega_y + \alpha_z & -\omega_x^2 - \omega_z^2 & \omega_y\omega_z - \alpha_x \\ a_z - g_z & \omega_x\omega_z - \alpha_y & \omega_y\omega_z + \alpha_x & -\omega_y^2 - \omega_x^2 \\ 0 & 0 & a_z - g_z & g_y - a_y \\ 0 & g_z - a_z & 0 & a_x - g_x \\ 0 & a_y - g_y & g_x - a_x & 0 \end{pmatrix}, \tag{4.9}$$

$$V_I = \begin{pmatrix} 0 & 0 & 0 & 0 & 0 & 0 \\ 0 & 0 & 0 & 0 & 0 & 0 \\ 0 & 0 & 0 & 0 & 0 & 0 \\ \alpha_x & \alpha_y - \omega_x\omega_z & \alpha_z + \omega_x\omega_y & -\omega_y\omega_z & \omega_y^2 - \omega_z^2 & \omega_y\omega_z \\ \omega_x\omega_z & \alpha_x + \omega_y\omega_z & \omega_z^2 - \omega_x^2 & \alpha_y & \alpha_z - \omega_x\omega_y & -\omega_x\omega_z \\ -\omega_x\omega_y & \omega_x^2 - \omega_y^2 & \alpha_x - \omega_y\omega_z & \omega_x\omega_y & \alpha_y + \omega_x\omega_z & \alpha_z \end{pmatrix}. \tag{4.10}$$

With Eq. 4.7 the inertial parameters can be estimated using the sensor values measured during the transfer motion of the robot. As the sensor data is updated in every control cycle, a recursive estimation technique has to be employed. A weighted recursive instrumental variables technique [46, 52] is applied, which combines signals of several sensors. These sensors are, as already mentioned above, a wrist-mounted inertial measurement unit, providing angular velocity signals and a wrist-mounted force-torque sensor, providing forces, torques and linear and angular accelerations. Additionally the encoder signals can be used as complementary measurements for angular velocity, linear acceleration and angular acceleration.

To successfully estimate the inertial parameters and simultaneously move the object from the start to the goal pose, a suitable trajectory has to be used. This trajectory should fulfill the requirement that all needed parameters for pose estimation are excited sufficiently. For example, a linear trajectory would not allow for estimation of the elements of the inertia matrix I. Therefore, the trajectory from start to goal is superimposed by sinusoidal movements in the three hand joints. To generate a measure of the quality of the pose estimation, the correlation matrix Ψ is used which consists of M different V matrices, so M successive measurements of all sensors, compiled together:

$$\Psi = \tilde{V}^T \tilde{V} \tag{4.11}$$

$$\tilde{V} = \begin{bmatrix} V_1^T V_2^T V_3^T \dots V_M^T \end{bmatrix} \tag{4.12}$$

The condition number of Ψ increases with increasing sensitivity of $^S\varphi$ to errors in $\tilde{V}$ and can therefore be used to observe the quality of the measurements. Furthermore, it has to be observed, whether the current estimate of the inertial parameters has already converged. This has the advantage that the non-time-optimal estimation trajectory can be stopped and a time-optimal place trajectory can be started as early as possible. The ground truth of the inertial parameters of the object is known (e.g. derived from the CAD model) and can be used to stop the parameter estimation procedure. To compare the ground truth and the measured parameters, it is essential to have rotational and translational invariant features. The only features usable in this context are the mass m and the principal moments of inertia. These moments can be calculated as the eigenvalues I_1, I_2 and I_3 of I and build the matrix $I_p = \mathrm{diag}(I_1, I_2, I_3)$. A four dimensional feature set $f = [m, I_1, I_2, I_3]$ can be built, using these features. The difference between the measured feature set fo and the ground truth feature set f_Q is estimated using the symmetric Kullback-Leibler divergence (SKLD) [96]. The SKLD J_{KL} of the two feature sets can be regarded as the distance between two probability distributions and takes the covariance Σ_{f_o} of the estimated feature set fo into account. fo and Σ_{f_o} can be obtained from the estimated parameter vector $^S\hat{\varphi}$ and its covariance matrix $\Sigma_{\hat{\varphi}}$ using the scaled unscented transform [40].

$$J_{KL} = \frac{1}{2}\left[\Delta f^T\left(\Sigma_{f_Q}^{-1} + \Sigma_{f_Q}^{-1}\right)\Delta f + tr\left(\Sigma_{f_Q}^{-1}\Sigma_{fo} + \Sigma_{fo}^{-1}\Sigma_{f_Q} - 2E\right)\right] \quad (4.13)$$

Here, $\Delta f = fo - f_Q$ and tr denotes the trace of a matrix. When J_{KL} falls below a predetermined threshold, the switch from the estimation trajectory to the time optimal direct place trajectory can be performed.

The gripper is attached to the force-torque sensor. Therefore, the sensors measure the forces and torques of the grasped objects as well as the gripper itself. To overcome this, the measurements resulting from the gripper have to be compensated. I.e., the inertial parameter vector of the gripper $^S I_{gripper}$ has to be subtracted from the estimated parameter vector $^S\hat{I}_{total}$ to yield an estimate of the inertial parameters of the object $^S\hat{I}_{obj}$.

$$^S\hat{I}_{obj} = {}^S\hat{I}_{total} - {}^S I_{gripper} \quad (4.14)$$

Further practical challenges, such as the elimination of sensor offsets, etc., have to be addressed to obtain robust and reliable estimates [47].

Using the estimated inertial parameters of the object, the pose of the object has to be derived. This is done in two steps. At first the coordinates of the center of mass (COM) are computed and then the principal axes of inertia. The center of mass is easily derived by using the parameters 2–4 of $^S\varphi$ and dividing by m. In order to compute the principal axes of inertia, the inertia matrix $^S I_{obj}$ is expressed with respect to the COM using the parallel-axis theorem [17]. The eigenvectors of the resulting matrix $^{COM}I_{obj}$ constitute the principal axes of inertia computed by eigen decomposition:

$$^{COM}I_{obj} = {}^S R\, I_p\, {}^S R^T \quad (4.15)$$

The columns of $^S\!R$ are the eigenvectors and I_p is a diagonal 3×3-matrix containing the eigenvalues which constitute the principal moments of inertia. The matrix of eigenvectors constitutes a rotation matrix $^S\!R$ relating the orientation of the principal axes to the sensor frame S. Therefore, a pose hypothesis $^S\!P$ can be composed of $^S\!R$ and a translation matrix given by $^S\!T = Trans\left(^S\!c_x, ^S\!c_y, ^S\!c_z\right)$.

$$^S\!P = ^S\!T\ ^S\!R\ M \tag{4.16}$$

The matrix M may denote the identity matrix E or any of the following rotations: $Rot\,(x, \pi)$, $Rot\,(y, \pi)$, and $Rot\,(z, \pi)$. This pose ambiguity results from the ambiguity of the principal axes of inertia.

This ambiguity problem can easily be addressed by using data of the depth camera used for gripper pose estimation or the geometry of the gripper itself. During the transfer motion simultaneous with the parameter estimation, simple visual features can be analyzed to delete the ambiguities. Figure 4.7a shows an example where a simple analysis of a local patch around the gripper pose is enough to solve the ambiguity problem. A set of pixels can be analyzed for depth differences to choose the correct option of the two orientations. In Fig. 4.7b the gripper geometry only allows for one grasp. Therefore only one of the two possible grasp poses is geometrically feasible. A simple geometric test, checking for collisions of the gripper fingers transformed relative to the CAD model can solve this problem.

When ambiguities arise and the visual features are not robust as well, situations may arise in which ambiguities are not resolvable in practice. Figure 4.7c shows an example in which the asymmetry around the symmetry axis of a cylindrical part neither causes robustly measurable differences during force/torque analysis nor can robustly be solved by depth image analysis because the height difference may be smaller then the sensor noise. In these cases, if the unknown rotation is important, a further sensor like a smart camera can be used to solve the problem.

(a)   **(b)**   **(c)**

Fig. 4.7 Ambiguities and ambiguity elimination. **a** The two depicted poses of the piston rod cannot be distinguished based on the inertial features. Simple clues provided by the vision system, however, can easily resolve the ambiguity as the pose hypotheses differ considerably. **b** The pose hypothesis associated with the *green* piston rod can be discarded. The gripper geometry and the maximum jaw distance render the depicted grasp impossible. **c** The pose hypotheses cannot be distinguished *in practice* because the object is nearly rotationally symmetric and therefore even minor estimation errors will lead to significant errors in the computation of the principal axes. However, image features may resolve such ambiguities as well. Graphic taken from [2]

4.4 Experimental Depth Map Based Bin-Picking

Two series of experiments, as described in the following subsections, were performed. In the first series, the ability of the system to grasp unknown objects was investigated. For this purpose, several different objects were picked and placed in a bin. The second series served to evaluate the applicability of the system as a bin-picking solution. With the aim of emulating industrial production environments, industrial metal parts in a bin were grasped and placed at a desired pose. In the first series, outer grasps were used whereas in the second series, inner grasps were employed as well.[4]

4.4.1 Hardware

For the experimental evaluation of the described approach, nearly the same setup as the one described in Sect. 3.3 was used. The only difference to this setup was the sensor. Instead of the SICK IVP Ruler, the Microsoft Kinect Sensor was employed (the complete hardware setup with all vision sensors can be seen in Fig. 5.21). This sensor was chosen for its data acquisition time of 30 fps or one thirtieth of a second. When a perspective camera like the Kinect is used, some modifications have to be made to the algorithms.

The Kinect is a depth camera. This means that every pixel of the acquired depth image stores the distance of the focal point to a point in the scene. This results in varying pixel sizes. Areas of the scene that are far away from the camera map to fewer pixels of the camera than areas that are near to the camera. So, surface patches mapped to pixels in the image are larger if they are farther away from the camera. To resolve this problem, an approximate pixel size is calculated of the nearest pixel in the image by using intrinsic camera parameters and the intercept theorem. The filter kernel K_G used for pick pose hypotheses generation has to be adapted to the pixel size. The kernel is therefore scaled by a factor s computed like described above, $K'_G = s \cdot K_G$. The convolution is then performed using K'_G.

As the Kinect sensor was not developed for industrial applications, where absolute accuracy is important, no effort was made to increase the stability of measurements w.r.t. thermal conditions. Due to this, a significant drift of sensor values can be observed when the temperature of the Kinect changes. In [20] this is analyzed in detail. To overcome this issue, 3 calibration markers were placed in the workspace with known coordinates. After a pick pose was generated, the coordinates of the three markers in the sensor image were measured. With these 3 correspondences, the z-coordinate of the pick pose was corrected.

Other hardware modifications to the setup of the prior experiments were the following. The 6D force-torque as well as 6D acceleration measurements were provided by a wrist-mounted sensor manufactured by JR3 (85M35A3-I40-D 200N12).

---

[4]Grasps in which the gripper is closed to grasp are called outer grasps, grasps in which the gripper is opened to grasp are called inner grasps.

(a) (b)

(c) (d)

Fig. 4.8 Four example test piles used in the experiments. The bin is filled with **a** identical metal parts, **b** mixed metal parts, **c** plastic toys and wooden toys, and **d** a mix of all available objects. Images taken from [2]

Angular velocity measurements were supplied by an Analog Devices IMU[5] (ADIS 16364). The inertial parameter estimation was performed on a PC running the QNX Neutrino real-time operating system. A distributed real-time middleware [21] enabled efficient communication between the computing nodes.

4.4.2 Grasping Unknown Objects

To analyze the performance of the system with unknown objects, several piles of objects in the work space were prepared (examples can be seen in Fig. 4.8). As test objects raw metal parts as well as plastic and wooden objects of arbitrary shapes were used. As the geometry of the objects was unknown, no grasp pose was estimated and the objects were only picked and dropped into another bin afterwards. A series of 261 grasp attempts has been performed. In contrast to the experiments presented in the following section, force sensing was merely employed to detect collisions. The data

---

[5]Inertia Measurement Unit.

acquisition time was $n/30$ s where n is the number of averaged Kinect scans. In our experiments, n is set to 3. The averaging was performed to reduce the sensor noise of the depth images. The time overhead did not increase the cycle times. The vision processing time, i.e. the time until a valid pick pose became available, was 18 ms (no notable variance). This value comprises 10 ms for convolution and maxima search as well as 8 ms for maxima evaluation including the computation of the approach distance and the orientation of the gripper. The number of skeleton pixels used for orientation estimation was around 20–30, depending on the scaling factor s, which resulted in a very fast evaluation. Obviously, both, the image acquisition and the vision processing (in total requiring 118 ms), could be executed during the place movement of the manipulator. With the test setup, the overall cycle time was 8 s from pick to pick, which is exactly the duration of the robot motion. The dynamics of the robot were reduced significantly since a pneumatic load limiter was used to prevent damages of the manipulator in case of collisions. However, this load limiter was also triggered by inertial forces and associated torques.

The total grasp success rate was 95.4 %. Unsuccessful grasps were caused by sensor noise and may be classified into two categories. The computed approach distance may be too small to allow for a stable grasp. In this case, the objects slipped off the gripper during the depart motion. Sensor noise may also result in minor collisions of the gripper with the object thus compromising the grasp. Due to the short acquisition and vision processing times, a new pick pose could be computed without significant time overhead in these rare cases, after the robot left the workspace above the bin. The average cycle time was therefore not increased notably.

4.4.3 Bin-Picking

The task of bin-picking, besides the picking of objects, also includes a defined placement. Simply dropping the grasped objects into another bin is not enough in most cases. In the experiments above, the placement was omitted to analyze the applicability of the gripper pose estimation. To integrate the gripper pose estimation technique into a bin-picking station, the grasp pose estimation, described in Sect. 4.3 is added to enable defined placement of the grasped objects. The sequence diagram of the bin-picking procedure can be seen in Fig. 4.9. This diagram shows that each computational step is performed directly before its result becomes important. All steps are computed simultaneous to the robot's movements. The gripper pose estimation starts when the robot leaves the workspace; the grasp pose estimation starts simultaneously with the place trajectory execution. The only time overhead occurs due to the superposed sinusoidal movements in the hand joints. This time overhead is approx. 20 % of the optimal transfer motion from bin to the feeder. However, as the estimation trajectory is canceled as soon as a sufficient pose estimate is available, the effective increase in the duration of the transfer motion is lower.

To show the applicability of the approach, a bin was filled with piston rods. Using the gripper pose estimation for inner grasps, the piston rods were picked up using

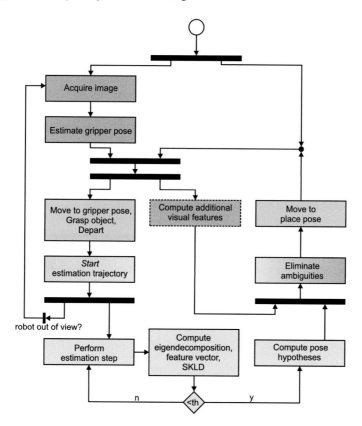

Fig. 4.9 Sequence diagram of the depth map based bin-picking system. The diagram reveals the high amount of possible parallel computations. The *red color* denotes the vision analysis, *green* denotes the force/torque/acceleration value analysis, and *blue* are the robot's movements. Diagram taken from [2]

their big holes and then placed in the feeder. As the inside grasp eliminates all but two DoFs of the object, which can be estimated unambiguously, no additional visual features are required. The key parameter is the angle of the piston rod around the approach vector of the gripper. The average absolute error obtained in 50 trials is 2.7°. This error enables a safe placement in the feeder as it can compensate minor angular deviations. A significant fraction of the estimation error is due to structural oscillations of the manipulator during the estimation trajectory. This problem is further aggravated by the load limiter.

4.5   Discussion

This chapter describes an extremely efficient approach to solve the bin-picking problem for arbitrary objects. It outperforms known approaches in terms of robustness
and cycle times. The good performance results from an efficient information man-
agement and data acquisition and analysis. 3D point clouds are avoided as a 2D
representation of the same data proves to be sufficient to solve the task. All this
makes it possible to use cheap 3D sensors whereas former approaches needed very
accurate and expensive vision hardware.

The disadvantage of this approach is that grasp poses cannot be restricted. If
certain parts are not allowed to be grasped at certain areas, this approach cannot be
applied. Furthermore, a set of sensors has to be added to the system to acquire the
inertial parameters needed for grasp pose estimation.

As grasping is not based on the model data of the object, it can also be used to pick
unknown objects. And even in this field of research, the performance is better than
known approaches. In these scenarios, when unknown objects have to be picked, a
defined placement is not possible.

Chapter 5
Normal Map Based Pose Estimation

The last section described an approach to solve the bin-picking problem using depth maps. These depth maps can be generated with special depth cameras or 3D scanners. To base an approach on 3D data has therefore at least one of the following disadvantages. Either the scanning device is expensive (state-of-the-art commercial laser scanners or structured light scanners) or a time consuming procedure has to be performed to acquire 3D coordinates or the resolution and the accuracy are very low (Kinect, ToF cameras) or a combination of these. Another disadvantage, in regard to the Kinect sensor, is the fixed hardware setup that results in a fixed measurement range. Many—and especially cheap sensors—only exist in a finite number of variations (e.g. the Kinect sensor is only available in a fixed hardware setup) and may therefore not be applicable to certain situations. Regarding service robotics, a further disadvantage may be the use of laser light which might not be regarded as eye-safe for use by humans.

As the described methods of this work, presented in the former sections, already reduce the costs of the sensors, not only in the monetary but also in the temporal sense, these considerations shall be continued within this chapter. Through the experiments of the last section, it turned out that using single shot data acquisition may result in superior performance of a system. So, the used scanning device should only need fractions of a second to acquire data.

Reviewing fast visual data acquisition techniques found in literature showed that it is possible to acquire *surface normal maps*[1] with a single camera shot.

---

[1] In this document, surface normal maps will be named simply "normal maps" for sake of simplicity.

© Springer International Publishing Switzerland 2016
D. Buchholz, *Bin-Picking*, Studies in Systems, Decision and Control 44,
DOI 10.1007/978-3-319-26500-1_5

5.1   The Normal Map

A normal map is an image of a scene, in which each pixel contains the normal of the surface intersected by the according viewing ray of that pixel. In contrast to depth maps, the distance between the focal point of the camera to the surface along this viewing ray is unknown. As each (normalized) normal has two degrees of freedom, normal maps can be color coded using e.g. the red color channel for the polar angle and the blue color channel for the azimuthal angle or three color channels for the three Cartesian coordinates of the normal (the latter is used within this work). With such a color representation, normal maps can be intuitively visualized.

The acquisition of normal maps is introduced briefly in the appendix in Sect. A.2 since it is not an essential contribution of this thesis.

As normal maps do not contain depth information but, in a mathematical sense the derivative of the depth, they are usually integrated, to generate 3D data, which are then used for various purposes. This, in contrast to the single shot acquisition, is time consuming and many assumptions, like smoothness constraints, and prior knowledge are needed for accurate depth estimation. For this reason, normal maps are rarely used for pose estimation.

But, as using single shot measurements is such a big advantage, and the integration of the normals to estimate depth is often not accurate enough, the following sections propose new approaches to directly use normal maps as basis for pose estimation.

5.2   Generic Pose Estimation Using Normal Maps

The problem with using normal maps for pose estimation of objects in 3D is that no explicit 3D data is stored in these maps, which is obviously the main reason not to use them for 3D pose estimation. But, with a calibrated camera, 3D viewing rays can be constructed through pixels with known normals. This is a very important aspect of the pose estimation technique presented here.

Based on CAD models, like in the first two chapters, a comparable representation of the model and the normal map has to be found first. The only quantities present in both data sets are surface normals. As it is not known, which of the normals in the image correspond to which normals of the model and no orientation information or position information of the object in the image is known, only the normals, without any positional information can be used. As the normal vectors only contain orientation information, the extraction of the normals of the image and the model separates the orientation estimation from the translation estimation leading to a two step pose estimation. Therefore, at first, the orientation is estimated using the extracted normals, and with a known orientation the translation is computed resulting in object poses. The two steps are presented separately in the next two sections. Each of the steps can be used without the other for further applications.

5.2.1 Orientation Estimation

As described in the introduction, the orientation information has to be extracted from the normal data. In other words, a comparable data representation has to be found for the image and the model.

This can be done by computing Extended Gaussian Images (EGI) [32] of both, the model and the normal map. An EGI can be interpreted as spherical histogram of normals. At each point of the sphere, the area of the object with the related normal is stored. The spherical coordinates φ and θ on the sphere of a normalized normal $\vec{n} = [n_x, n_y, n_z]^T$ can be computed as

$$\varphi = arccos\,(z)\,, \tag{5.1}$$

$$\theta = arctan\left(\frac{y}{x}\right). \tag{5.2}$$

Here, "area" means the sum of the sizes of the faces with corresponding normals, when dealing with the CAD model and the number of the pixels with corresponding normals when dealing with the normal map. Further, as tilted surface patches projected onto one pixel are larger than surface patches parallel to the image plane, each normal of the normal map contributes an area of

$$A(\boldsymbol{n}_i) = \frac{A_{\text{pixel}}}{\langle \boldsymbol{n}_i, \boldsymbol{r} \rangle} \tag{5.3}$$

to the EGI at the position of the normal $\boldsymbol{n}_i$. The sum of all $A(\boldsymbol{n}_i)$ then forms the complete EGI. Here, $\boldsymbol{n}_i$ is the normal, $\boldsymbol{r}$ the 3D viewing ray of the pixel, and A_{pixel} is the size of a single pixel. Of course, the distance of the surface patch does also effect the size of the area that is observed by one pixel. But, as the distance is unknown and, under consideration of a weak perspective (see Sect. A.2.1) does not vary very much within the model, the effect of the distance is ignored.

The two EGIs each represent a function on the unit sphere (see Fig. 5.1). A successful orientation estimation is the comparison of both spheres and the search for an optimal overlap of both functions. Due to the monocular image acquisition, the EGI of the normal map is subject of shadowing effects. Considering a convex object, only the upper half of it is visible and thus, only the upper half equals the model's EGI. If concave objects are used, even more shadowing effects are present. Assuming a perfect rotational alignment between model and image, this can be expressed as

$$E_{\text{model}}(\boldsymbol{\omega}) = s\left(E_{image}(\boldsymbol{\omega}) + h(\boldsymbol{\omega})\right) + l(\boldsymbol{\omega}). \tag{5.4}$$

Here, $E_{\text{model}}(\boldsymbol{\omega})$ and $E_{image}(\boldsymbol{\omega})$ are the two EGIs, s is a scaling factor, $h(\boldsymbol{\omega})$ is the hidden part of the upper hemisphere and $l(\boldsymbol{\omega})$ is the lower hemisphere of the EGI of the object. $\boldsymbol{\omega} = (\varphi, \theta)^T$ is the spherical coordinate.

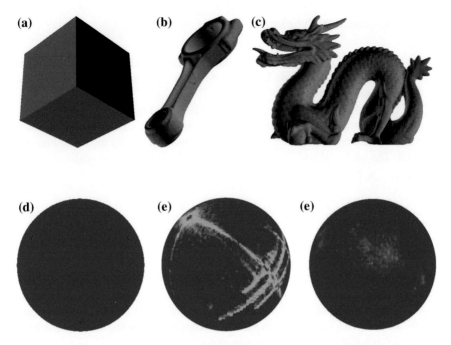

Fig. 5.1 Extended Gaussian Images. The values on the spheres are scaled. *Red* are the highest values and *blue* the lowest values. **a, d** Cube model and EGI. **b, e** Piston rod model and EGI. **c, f** Dragon model and EGI

To find the optimal orientation of the two EGIs, a correlation of the two spherical functions has to be computed. This correlation determines the three rotational DOFs between the two EGIs. The mathematical background of this is the topic of the next section.

Mathematical Background of the Orientation Estimation

The comparison of two functions can be computed by a correlation of these functions. The location of the maximum of the correlation function describes the amount of "shift" between the two functions that needs to be applied for maximum overlap. As the functions present here are defined on the sphere $\mathbb{S}^2$, a detailed introduction on the mathematical theory is given here. In image analysis, the correlation of functions defined on the plane $\mathbb{R}^2$ is very popular. Therefore, parallels are always mentioned to clarify the theory. At first, the continuous case is described and the results are then transferred to discrete spheres. At first, Extended Gaussian Images, the sphere $\mathbb{S}^2$ and the rotation group $SO(3)$ are introduced in detail. Then, an efficient way of correlation of functions on the sphere is introduced. At last, the energy analysis of spherical functions and its application in the context of pose estimation is shown.

The mathematical background presented in this chapter is mainly derived from [45, 86].

Extended Gaussian Images, the Sphere $\mathbb{S}^2$ and the Rotation Group $SO(3)$. To simplify the complex problem of pose estimation using normal maps, it was split into two sub-problems. To decouple the orientation estimation and the translation estimation, all normals were mapped onto the unit sphere using the according face size as weight in a first step. By this, all available normals were combined, producing two so called Extended Gaussian Images (EGI). The EGI usually is used in case of 3D point clouds and only for coarse registration prior to an ICP algorithm [55]. To stay in mathematical terms, we consider the two EGIs as real valued functions $f : \mathbb{S}^2 \to \mathbb{R}$ and $g : \mathbb{S}^2 \to \mathbb{R}$ defined on the unit sphere $\mathbb{S}^2$ embedded into $\mathbb{R}^3$, i.e.

$$\mathbb{S}^2 = \left\{ \boldsymbol{x} \in \mathbb{R}^3 \mid \|\boldsymbol{x}\| = 1 \right\}. \tag{5.5}$$

Each coordinate $\boldsymbol{x}$ on the unit sphere can be transformed from spherical coordinates (the azimuthal angle φ and the polar angle θ) to Cartesian coordinates by

$$\boldsymbol{x}(\varphi, \theta) = (x_1, x_2, x_3)^T = \begin{pmatrix} \sin(\theta)\cos(\varphi) \\ \sin(\theta)\sin(\varphi) \\ \cos(\theta) \end{pmatrix} ; \varphi \in [0, 2\pi) ; \theta \in [0, \pi]. \tag{5.6}$$

To rotate two different points onto each other, the rotation matrix $\boldsymbol{R} \in \mathbb{R}^{3\times3}$ can be used. The matrix $\boldsymbol{R}$ fulfills $det(\boldsymbol{R}) = 1$ and $\boldsymbol{R}^T \boldsymbol{R} = \boldsymbol{I}$ with $\boldsymbol{I}$ being the identity matrix. The group that consists of all rotation matrices $\boldsymbol{R}$ is called Rotation Group $SO(3)$. The rotation of a point $\boldsymbol{v} \in \mathbb{S}^2$ can be computed as

$$\boldsymbol{v}_R = \boldsymbol{R}\boldsymbol{v}; \boldsymbol{R} \in SO(3). \tag{5.7}$$

There are different ways of rotation parametrization for the elements of $SO(3)$. For the techniques presented here, the Euler angle parametrization is used. Every rotation matrix $\boldsymbol{R}$ can be constructed using three angles α, β and γ with $\alpha, \gamma \in [0, 2\pi)$ and $\beta \in [0, \pi]$. $\boldsymbol{R}(\alpha, \beta, \gamma)$ is then defined as

$$\boldsymbol{R}(\alpha, \beta, \gamma) = \boldsymbol{R}_z(\alpha)\boldsymbol{R}_y(\beta)\boldsymbol{R}_z(\gamma) \tag{5.8}$$

with

$$\boldsymbol{R}_z(\alpha) = \begin{pmatrix} \cos(\alpha) & -\sin(\alpha) & 0 \\ \sin(\alpha) & \cos(\alpha) & 0 \\ 0 & 0 & 1 \end{pmatrix} \tag{5.9}$$

and

$$\boldsymbol{R}_y(\beta) = \begin{pmatrix} \cos(\beta) & 0 & \sin(\beta) \\ 0 & 1 & 0 \\ -\sin(\beta) & 0 & \cos(\beta) \end{pmatrix}. \tag{5.10}$$

Since the two EGIs described in Sect. 5.2.1 are rotated versions of each other, at least theoretically,[2] the orientation estimation problem can be reformulated as finding the rotation matrix $R \in SO(3)$ such that $f(x) = g(Rx)$, where f, g are real valued functions representing the two EGIs. As f and g are defined on $\mathbb{S}^2$ and R is an element of $SO(3)$, the close relationship between the rotation group and the unit sphere has to be analyzed.

The defined rotation matrix R was parameterized using Euler angles. Another representation of rotation matrices is the axis-angle parametrization. Given R, its rotation axis r is defined by the eigenvector of R, w.r.t. the eigenvalue $\lambda = 1$, as

$$r = \frac{1}{\|v\|} v \text{ with } v = \begin{pmatrix} g_{23} - g_{32} \\ g_{31} - g_{13} \\ g_{12} - g_{21} \end{pmatrix} \text{ and } R = \left(g_{ij}\right)_{j,k=1,2,3} \in SO(3); R \neq I.$$

(5.11)

The rotation angle ω is defined as

$$\omega = \cos^{-1}\left(\frac{tr(R) - 1}{2}\right).$$

(5.12)

As $\|r\| = 1$, it follows that $r \in \mathbb{S}^2$ and the link from $SO(3)$ to $\mathbb{S}^2$ is obvious. When the axis r and the angle ω are given, R can be constructed using

$$R = R(\omega, n) = I + \sin(\omega N) + (1 - \cos(\omega)) N^2$$

(5.13)

with

$$n = (n_1, n_2, n_3)^T \text{ and } N = \begin{pmatrix} 0 & -n_3 & n_2 \\ n_3 & 0 & -n_1 \\ -n_2 & n_1 & 0 \end{pmatrix}.$$

(5.14)

This directly leads to a possible metric Φ for the rotation group [36]. This metric is defined as

$$\Phi(R_i, R_j) := |\Theta|, \text{ with } R(\Theta, n) = R_i R_j^T; R_i, R_j \in SO(3).$$

(5.15)

This metric is invariant towards rotations and is essential for the comparison of different rotations.

With these basics, the comparison of two spherical images $f, g : \mathbb{S}^2 \to \mathbb{C}$ is basically the same as between two planar images $h, i : \mathbb{R}^2 \to \mathbb{C}$. A cross correlation of f and g computes the similarity of two functions. On the sphere, this can be written as

$$(f \star g)(R) = \int_{\mathbb{S}^2} f(\omega)\overline{g(R^{-1}\omega)}d\omega; f, g : \mathbb{S}^2 \to \mathbb{C}.$$

(5.16)

---

[2] As already described in Sect. 5.2.1, the spheres are only approximately similar, due to shadowing effects, limited viewpoints and sensor noise.

The problem with the direct computation of the correlation in this way is that high correlation maxima are formed by high valued areas on f independent of the pattern on g. To enhance the robustness of the correlation and to suppress these errors, the normalized cross correlation

$$NC(R) = \frac{\int_{\mathbb{S}^2} \left(f\left(\omega\right) - \tilde{f}_{\mathcal{W}} \right) \left(g\left(R\omega\right) - \tilde{g}\right) d\omega}{\sqrt{\int_{\mathcal{W}} \left| f\left(\omega\right) - \tilde{f}_{\mathcal{W}} \right|^2 d\omega} \sqrt{\int_{\mathcal{W}} \left| g\left(\omega\right) - \tilde{g}\right|^2 d\omega}}, \tag{5.17}$$

cf. [78] can be used. Here, $\mathcal{W}$ is the window defined by the limited area of the EGI of the image, i.e., a hemisphere. $\tilde{f}_{\mathcal{W}}, \tilde{g} \in \mathbb{C}$ are the mean of f under the rotated hemisphere of the EGI of the image and the mean of the EGI of the image g, respectively.

Correlation of Functions on $\mathbb{S}^2$. The evaluation of the Eqs. (5.16) and (5.17) is computationally very costly. But, as known from image analysis, the convolutions can efficiently be computed using the fast Fourier Transform. This is also true for functions defined on the sphere. To describe the mathematical background of the efficient rotation estimation based on unit spheres, at first an introduction to the harmonic analysis of SO(3) is needed. As $\mathbb{S}^2$ is embedded into $SO(3)$, the introduction of analysis of functions on the rotation group and the unit sphere are important.

The integration of a function $f : \mathbb{S}^2 \to \mathbb{C}$, depending on spherical coordinates $\xi = \xi(\alpha, \beta) \in \mathbb{S}^2$ is defined as

$$\int_{\mathbb{S}^2} f(\xi)d\xi = \frac{1}{4\pi} \int_0^{2\pi} \int_0^\pi f(\xi) \sin(\beta)d\alpha d\beta \tag{5.18}$$

With this definition and the restrictions of the functions to be square integrable $f \in L^2(\mathbb{S}^2)$ and to have finite energy $\int_{\mathbb{S}^2} f(\xi)\overline{f(\xi)}d\xi < \infty$, the inner product of two functions $f, g \in \mathbb{S}^2$ can be defined as

$$\langle f, g \rangle = \int_{\mathbb{S}^2} f(\xi)\overline{g(\xi)}d\xi \tag{5.19}$$

while the convolution of two functions $f, g \in \mathbb{S}^2$ is given by

$$(f, g)(R) = \int_{\mathbb{S}^2} f(\xi)g(R^{-1}\xi)d\xi. \tag{5.20}$$

The idea of the Fourier transform is, to describe a function as sum of weighted basis functions. On the line $\mathbb{R}$, which is very popular with all engineers, these functions are sines and cosines. And every function $f : \mathbb{R} \to \mathbb{C}$ can be described as

$$f(x) = \int_{\mathbb{R}} \hat{f}(u)e^{2\pi iux}du. \tag{5.21}$$

The term $e^{2\pi i u x}$ describes the mentioned sines and cosines and therefore the basis for functions on $\mathbb{R}$.

To transfer the Fourier transform to the sphere, at first an orthogonal basis for $L^2(\mathbb{S}^2)$ has to be found. This basis are the well known spherical harmonics with degree $l \in \mathbb{N}_0$ and order $m = -l, \ldots l$, which are defined as

$$Y_l^m(\xi) = \sqrt{\frac{2l+1}{4\pi}} \sqrt{\frac{(l-m)!}{(l+m)!}} P_l^m(\cos(\theta)) e^{im\varphi}. \qquad (5.22)$$

Here, $\xi \in \mathbb{S}^2$ with coordinates $(\varphi, \theta) \in [0, 2\pi) \times [0, \pi]$ and $P_l^m : [-1, 1] \rightarrow \mathbb{R}$ are associated Legendre Polynomials

$$P_l^m(x) = (-1)^m (1-x^2)^{\frac{m}{2}} \frac{d^m}{dx^m} P_l(x) = \frac{(-1)^m}{2^l l!} (1-x^2)^{\frac{m}{2}} \frac{d^{l+m}}{dx^{l+m}} (x^2-1)^l \quad (5.23)$$

that are constructed as derivatives of ordinary Legendre polynomials $P_l(x)$. The spherical harmonics are the eigenfunctions of the Laplacian on the sphere. Furthermore, the spherical harmonics satisfy the orthogonality relation

$$\int_{\mathbb{S}^2} Y_l^m(\xi) \overline{Y_{l'}^{m'}(\xi)} d\xi = \delta_{ll'} \delta_{mm'}. \qquad (5.24)$$

By δ_{ab} in this context, the Kroenecker delta

$$\delta_{ab} = \begin{cases} 1, & \text{if } a = b \\ 0, & \text{otherwise} \end{cases} \qquad (5.25)$$

is denoted, which shall not be mistaken for a Dirac delta function.

The subspace $\text{Harm}_l(\mathbb{S}^2) = \text{span}\left\{Y_l^m | m = -l, \ldots, l\right\}$ spanned by spherical harmonics with a fixed degree $l \in \mathbb{N}$ is called harmonic space of degree l. The harmonic spaces $\text{Harm}_l(\mathbb{S}^2)$ provide a complete system of $SO(3)$-invariant subspaces of $L^2(\mathbb{S}^2)$, i.e.,

$$L^2(\mathbb{S}^2) = \text{clos}_{L^2} \bigoplus_{l=0}^{\infty} \text{Harm}_l(\mathbb{S}^2). \qquad (5.26)$$

Therefore, the spherical harmonics can be used as basis of $L^2(\mathbb{S}^2)$. In the following, the harmonic subspaces Harm_l are also called band spaces $\mathcal{B}_l$.

Using this basis, the Fourier expansion of functions $f : \mathbb{S}^2 \rightarrow \mathbb{C}$ can be computed. As the basis functions have the degree l and in each *band* defined by a certain degree l' there are $2l' + 1$ basis functions of order $m' = -l', \ldots, l'$, this expansion can be constructed by

$$f(\xi) = \sum_{l=0}^{\infty} \sum_{m=-l}^{l} \hat{f}_l^m Y_l^m(\xi), \quad \xi \in \mathbb{S}^2. \qquad (5.27)$$

The $\hat{f}_l^m$ are called Fourier coefficients and can be computed as

$$\hat{f}_l^m = \langle f, Y_l^m \rangle. \tag{5.28}$$

The appearance of the spherical harmonics can be seen in Fig. 5.2.

Now with a basis for functions on $\mathbb{S}^2$, the concept of (5.7) can be transferred to complete functions. As already said, the EGIs only differ in a relative rotation. Therefore, the relation between the two oriented functions on the sphere can be written as

$$C(\boldsymbol{R}) = \int_{\mathbb{S}^2} f(\boldsymbol{\xi})\overline{g(\boldsymbol{R}^{-1}\boldsymbol{\xi})}d\boldsymbol{\xi}. \tag{5.29}$$

This is, of course, the correlation of the two functions, both defined on the sphere. The maximum of the function C then is the optimal solution to the orientation estimation problem. By transferring this problem into the Fourier domain, using the tools described above, the convolution simplifies to a multiplication just like it does in the plane. At first, the functions f and g are expanded in terms of spherical harmonics

$$f(\boldsymbol{\xi}) = \sum_{l=0}^{\infty}\sum_{m=-l}^{l} \hat{f}_l^m Y_l^m(\boldsymbol{\xi}) \text{ and } g(\boldsymbol{\xi}) = \sum_{l'=0}^{\infty}\sum_{n=-l'}^{l'} \hat{g}_{l'}^n Y_{l'}^n(\boldsymbol{\xi}). \tag{5.30}$$

Combining formula (5.29) and (5.27) leads to

$$C(\boldsymbol{R}) = \int_{\mathbb{S}^2} \left[\sum_{l=0}^{\infty}\sum_{m=-l}^{l} \hat{f}_l^m Y_l^m(\boldsymbol{\xi}) \right] \overline{\left[\Lambda(\boldsymbol{R}) \sum_{l'=0}^{\infty}\sum_{n=-l'}^{l'} \hat{g}_{l'}^n Y_{l'}^n(\boldsymbol{\xi}) \right]} d\boldsymbol{\xi}. \tag{5.31}$$

At this point, a property of the spherical harmonics can be exploited to further simplify the formula. This property is that a rotated spherical harmonic always stays in its band, i.e., its frequency is rotational invariant. This expresses as

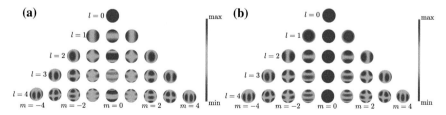

Fig. 5.2 Visualization of the spherical harmonincs. **a** Real part of the spherical harmonics. **b** Imaginary part of the spherical harmonics. Image taken with permission from [49]

$$Y_l^n(\boldsymbol{R}^{-1}\boldsymbol{\xi}) = \sum_{m=-l}^{l} Y_l^m(\boldsymbol{\xi}) D_l^{m,n}(\boldsymbol{R}). \qquad (5.32)$$

This property is called the representation property.

By $D_l^{m,n}$, the Wigner-D functions are denoted. A full derivation of these functions can be found in [86]. Explicitly, they can be written w.r.t. their Euler angles as

$$D_l^{m,n}(\boldsymbol{R}(\alpha, \beta, \gamma)) = e^{-im\alpha} e^{-in\gamma} d_l^{m,n}(\cos(\beta)). \qquad (5.33)$$

The term $d_l^{m,n}(x)$ are the so called Wigner-d functions that are defined as

$$d_l^{m,n}(x) = \frac{(-1)^{l-n}}{2^l} \sqrt{\frac{(l+m)!}{(l-n)!(l+n)!(l-m)!}} \sqrt{\frac{(1-x)^{n-m}}{(1+x)^{m+n}}} \frac{d^{l-m}}{dx^{l-m}} \frac{(1-x)^{n+l}}{(1+x)^{n-l}}. \qquad (5.34)$$

The Wigner-D functions are a generalization of spherical harmonics, which gives a link from $SO(3)$ to $\mathbb{S}^2$. Consequently, the spherical harmonics can be expressed, using Wigner-D functions

$$\begin{aligned} Y_l^m(\alpha, \beta) &= \sqrt{\frac{2l+1}{4\pi}} e^{im\alpha} d_l^{m,0}(\cos(\beta)) \\ &= \sqrt{\frac{2l+1}{4\pi}} \overline{D_l^{m,0}(\boldsymbol{R}(\alpha, \beta, \gamma))} \\ &= \sqrt{\frac{2l+1}{4\pi}} D_l^{0,-m}(\boldsymbol{R}(\gamma, \beta, \alpha)). \end{aligned} \qquad (5.35)$$

The third Euler angle $\gamma \in [0, 2\pi)$ can be chosen freely because $\mathbb{S}^2$ only has the two parameters α and β.

Wigner-d functions are symmetric in certain ways

$$\begin{aligned} d_l^{m,n}(-x) &= (-1)^{l+n} d_l^{-m,n}(x) \text{ and} \\ d_l^{m,n}(x) &= (-1)^{m+n} d_l^{n,m}(x) \\ &= (-1)^{m+n} d_l^{-m,-n}(x) \\ &= d_l^{-n,-m}(x). \end{aligned} \qquad (5.36)$$

Now, combining (5.31) and (5.32) leads to

$$C(\boldsymbol{R}) = \sum_{l=0}^{\infty} \sum_{l'=0}^{\infty} \sum_{m=-l}^{l} \sum_{k=-l'}^{l'} \sum_{n=-l'}^{l'} \overline{D_{l'}^{k,n}(\boldsymbol{R})} \hat{f}_l^m \overline{\hat{g}_{l'}^n} \int_{\mathbb{S}^2} Y_l^m(\boldsymbol{\xi}) \overline{Y_{l'}^k(\boldsymbol{\xi})} d\boldsymbol{\xi}. \qquad (5.37)$$

Of the symmetry properties of Wigner-d functions (5.36) follows that $\overline{D_l^{m,n}} = D_l^{-m,-n}$. With this and the knowledge of the orthogonality of the spherical harmonics (5.24), Eq. (5.37) further simplifies to

$$
\begin{aligned}
C(\boldsymbol{R}) &= \sum_{l=0}^{\infty} \sum_{m=-l}^{l} \sum_{n=-l}^{l} (-1)^{m+n} \hat{f}_l^m \overline{\hat{g}_l^n} D_l^{-m,-n}(\boldsymbol{R}) \\
&= \sum_{l=0}^{\infty} \sum_{m=-l}^{l} \sum_{n=-l}^{l} (-1)^{m+n} \hat{f}_l^{-m} \overline{\hat{g}_l^{-n}} D_l^{m,n}(\boldsymbol{R}).
\end{aligned}
\tag{5.38}
$$

With this, the link between convolution of functions on the plane $\mathbb{R}^2$ and the sphere $\mathbb{S}^2$ is perfectly obvious. One can see in (5.38) that the correlation reduces to a multiplication using the Fourier transform. This immensely simplifies the computation of the cross correlation (Eq. (5.16)).

Further analysis of the Wigner-D functions shows that

$$
\left\langle D_l^{m,m'}, D_{l'}^{n,n'} \right\rangle = \frac{8\pi^2}{2l+1} \delta_{mn} \delta_{m'n'} \delta_{ll'}.
\tag{5.39}
$$

Therefore, the Wigner-D functions themselves are an orthogonal basis of $SO(3)$ and the $SO(3)$ Fourier series is

$$
f(\boldsymbol{R}) = \lim_{L \to \infty} \sum_{l=0}^{L} \sum_{m=-l}^{l} \sum_{n=-l}^{l} \hat{f}_l^{m,n} D_l^{m,n}(\boldsymbol{R}), \; f \in L^2(SO(3); \mathbb{C})
\tag{5.40}
$$

with

$$
\hat{f}_l^{m,n} = \frac{2l+1}{8\pi^2} \left\langle f, D_l^{m,n} \right\rangle.
\tag{5.41}
$$

Applying Spherical Harmonics to the Orientation Estimation Problem. The presented theory in the former section has to be applied and the presented algorithms have to be evaluated in an efficient and robust way. Therefore, some modifications are presented that allow the application of the mathematics to real world tasks. To enhance the efficiency of the normalized cross correlation (5.17) some more deliberations need to be made, as in Eq. (5.17) the additional normalization terms N_g and N_f are present

$$
N_g = \sqrt{\int_{\mathcal{W}} |g(\omega) - \tilde{g}|^2 \, d\omega} = \sqrt{\int_{\mathcal{W}} (g(\omega) - \tilde{g}) \overline{(g(\omega) - \tilde{g})} d\omega}
\tag{5.42}
$$

$$N_f(\mathbf{R}) = \sqrt{\int_{\mathcal{W}(\mathbf{R})} \left| f(\omega) - \tilde{f}_{\mathcal{W}} \right|^2 d\omega}$$

$$= \sqrt{\int_{\mathcal{W}(\mathbf{R})} \left(f(\omega) - \tilde{f}_{\mathcal{W}} \right) \overline{\left(f(\omega) - \tilde{f}_{\mathcal{W}} \right)} d\omega}, \qquad (5.43)$$

cf. [34]. To recap, $\mathcal{W}$ is the window defined by the limited area of the EGI of the image, i.e., a hemisphere and $\mathcal{W}(\mathbf{R})$ is its rotated version. N_g and N_f are the energies of the two functions subtracted by their means as the energy E_f of a function $f \in L^2(\mathbb{S}^2; \mathbb{C})$ is defined as

$$E_f = \int_{\mathbb{S}^2} f(\omega)\overline{f(\omega)} d\omega. \qquad (5.44)$$

Parseval's theorem gives a link between the Fourier coefficients of a function and the function itself. For the spherical case, this is

$$E_f = \int_{\mathbb{S}^2} f(\omega)\overline{f(\omega)} d\omega = \sum_{l=0}^{L} \sum_{m=-l}^{l} \left| \hat{f}_l^m \right|^2. \qquad (5.45)$$

The spherical harmonic with degree 0 and order 0 is

$$Y_0^0 \equiv \frac{1}{\sqrt{4\pi}}. \qquad (5.46)$$

Therefore, the mean of a function $f : \mathbb{S}^2 \to \mathbb{C}$ computes as

$$\tilde{f} = \frac{\int_{\mathbb{S}^2} f(\omega) d\omega}{\int_{\mathbb{S}^2} d\omega} = \frac{\int_{\mathbb{S}^2} f(\omega) \frac{1}{\sqrt{4\pi}} \sqrt{4\pi} d\omega}{4\pi} = \frac{\sqrt{4\pi} \langle f, Y_0^0 \rangle}{4\pi} = \frac{1}{\sqrt{4\pi}} \hat{f}_0^0. \qquad (5.47)$$

Using (5.45) and (5.47), Eq. (5.42) becomes:

$$N_g^2 = \sum_{l=0}^{L} \sum_{m=-l}^{l} \left| \hat{g}_l^m \right|^2 - \frac{1}{\sqrt{4\pi}} (\hat{g}_0^0)^2 = \left(\hat{g}_0^0 - \frac{\hat{g}_0^0}{\sqrt{4\pi}} \right)^2 + \sum_{l=1}^{L} \sum_{m=-l}^{l} \left| \hat{g}_l^m \right|^2. \qquad (5.48)$$

The coefficients $\hat{g}_l^m$ are also needed for the standard cross correlation. Therefore, the computation of N_g does not change the complexity of the overall algorithm. To efficiently compute the normalization term $N_f(\mathbf{R})$, the window function $W(\omega)$ is introduced

$$W(\omega) = \begin{cases} 1, & \text{if } \omega \in \mathcal{W} \\ 0, & \text{otherwise} \end{cases} \qquad (5.49)$$

to consider the rotational dependency of N_f of ω. Hence, (5.43) can be rewritten as

$$N_f^2(\mathbf{R}) = \int_{SO(3)} \left| f(\omega) - \tilde{f}_W \right|^2 W(\mathbf{R}^{-1}\omega) d\omega. \tag{5.50}$$

This corresponds to the cross correlation of the function $U(\omega) = \left| f(\omega) - \tilde{f}_W \right|^2$ with the window function W and is therefore computable using the fast Fourier transform. So, (5.43) finally simplifies to:

$$N_f(\mathbf{R}) = \sqrt{(U \star W)(\mathbf{R})} \tag{5.51}$$

As W is constant, $N_f(\mathbf{R})$ can be computed offline and only once per model. Therefore, the overall complexity of the normalized cross correlation is the same as the complexity of the cross correlation.

Energy Analysis of Functions on $\mathbb{S}^2$

With Eq. (5.4), the relation of the two EGIs, present in the orientation estimation process, can be described. When the "amount" of $h(\omega)$ and $l(\omega)$ becomes large in this equation, the similarity of the two EGIs may become quite small and the estimation might fail. This is mainly the case, when very complex, concave objects have to be located. In these scenarios, it is necessary to enhance the search process by a limitation of the search space.

The Fourier analysis, described in the former section, offers an additional feature which can be analyzed for this purpose. Introduced in Eq. (5.45), Parseval's theorem describes the link between a function, its spectrum, and its energy. Recall that the energy E_f of a function $f \in L^2(\mathbb{S}^2; \mathbb{C})$ can be computed as

$$E_f = \int_{\mathbb{S}^2} f(\omega)\overline{f(\omega)} d\omega = \sum_{l=0}^{L} \sum_{m=-l}^{l} \left| \hat{f}_l^m \right|^2. \tag{5.52}$$

Here, l is the degree of the according spherical harmonics. All spherical harmonics Y_l^m with fixed l build the lth band space $\mathcal{B}_l$. The union of all $\mathcal{B}_l$ builds $L^2(\mathbb{S}^2)$. And a function f can be represented by the union of functions $f_l \in \mathcal{B}_l$ by

$$f(\omega) = \sum_{n=1}^{\infty} f_l(\omega) \tag{5.53}$$

with

$$f_l(\omega) := \sum_{m=-l}^{l} \hat{f}_l^m Y_l^m(\omega). \tag{5.54}$$

The band energy of the band space $\mathcal{B}_l$ can be computed by

$$E_{f_l} = \sum_{m=-l}^{l} \left| \hat{f}_l^m \right|^2. \tag{5.55}$$

Due to the rotational invariance of the band spaces, a vector of all band energies can be used as rotational invariant feature vector $e_f := \{E_{f_0}, E_{f_1}, \ldots, E_{f_N}\}$.

In [41] this feature vector was used for a model search in a data base. For the orientation estimation, it can be used to reduce the search space. By rendering a set of normal maps of the object of equispaced poses on the surrounding sphere of the object, e_f can be computed for each rendered normal map and a data base can be built containing the virtual camera poses, the EGIs of the rendered normal maps, and the energy vectors as keys. When a new EGI of a scene is acquired, a data base search can be done to find the best fitting feature vector. By this, the nearest viewing direction can be found and the correlation of the two EGIs can be computed in which $h(\omega)$ and $l(\omega)$ are very small.

Another interesting aspect is that if the density of camera poses around the object is high, the nearest neighbor match of the two energy vectors only leaves one degree of freedom, as the image may only be rotated around the viewing direction. Interestingly, this degree of freedom can be computed during the translation estimation steps, as will be described in Sect. 5.2.2. An energy lookup table may therefore be enough for orientation estimation and the correlation of the two EGIs may be redundant. This aspect is a topic for further research.

Sample Sets for $\mathbb{S}^2$ and $SO(3)$

To numerically compute the correlations needed for pose estimation, the algorithms described above have to be applied to discrete sets. Therefore sample sets for two different manifolds, i.e., the unit sphere $\mathbb{S}^2$ (the EGIs) as well as the rotation group $SO(3)$ have to be found. Furthermore, the Fourier transforms have to be adapted to work on discrete sets of points. As $\mathbb{S}^2$ and $SO(3)$ are closely related, the sample sets are very similar.

The discrete Fourier series of a function $f : \mathbb{S}^2 \to \mathbb{C}$ with according spherical harmonics $Y_l^m | (l, m) \in \mathcal{I}_S(L)$, the set of indices $\mathcal{I}_S(L) = \{(l, m)|l \in \mathbb{N}_0, l \leq L, m \in \mathbb{Z} \wedge -l \leq m \leq l\}$, and bandwidth L is defined as

$$f = Y\hat{f}, \tag{5.56}$$

$$\hat{f} = w^T \overline{Y} f, \tag{5.57}$$

with

$$f = (f(s_1), \ldots, f(s_N)) \in \mathbb{C}^N,$$

$$\hat{f} = \left(\hat{f}_0^0, \hat{f}_1^{-1}, \hat{f}_1^0, \hat{f}_1^1, \ldots, \hat{f}_L^L \right) \in \mathbb{C}^{(L+1)^2},$$

$$Y = \left(Y_l^m(s_i) \right)_{s_i \in \mathcal{S};\ (l,m) \in \mathcal{I}_Y(L)} \in \mathbb{C}^{N \times (L+1)^2}, \tag{5.58}$$

$$w = (w_1, w_2, \ldots, w_N) \in \mathbb{R}^N.$$

For functions $f : SO(3) \to \mathbb{C}$ and according Wigner-D functions $D_l^{m,n}|(l, m, n) \in \mathcal{I}_R(L)$, the set of indices $\mathcal{I}_R(L) = \{(l, m, n) | l \in \mathbb{N}_0, l \leq L; m, n \in \mathbb{Z} \wedge -l \leq m, n \leq l\}$, and bandwidth L the Fourier series is similarly defined as

$$f = D\hat{f}, \tag{5.59}$$

$$\hat{f} = w^T \overline{D} f, \tag{5.60}$$

with

$$f = (f(s_1), \ldots, f(s_N)) \in \mathbb{C}^N,$$

$$\hat{f} = \left(\hat{f}_0^{0,0}, \hat{f}_1^{-1,-1}, \hat{f}_1^{-1,0}, \hat{f}_1^{1,1}, \hat{f}_1^{0,-1}, \ldots, \hat{f}_L^{L,L} \right) \in \mathbb{C}^{(L+1)(2L+1)(2L+3)},$$

$$D = \left(D_l^{m,n}(r_i) \right)_{r_i \in \mathcal{R};\ (l,m,n) \in \mathcal{I}_W(L)} \in \mathbb{C}^{N \times \frac{1}{3}(L+1)(2L+1)(2L+3)}, \tag{5.61}$$

$$w = (w_1, w_2, \ldots, w_N) \in \mathbb{R}^N.$$

The two sample sets $\mathcal{I}_S$ and $\mathcal{I}_R$ are essential for a successful computation. Therefore, they have to be chosen carefully. Different sample sets differ in their point distribution and their computational complexity.

Sample Sets for $\mathbb{S}^2$. EGIs were used to represent the rotational relation between the model and the normal map. All normals of both, the map and the model are stored on the spheres. Therefore, it is essential that on the one hand, it is possible to efficiently find the nearest available neighbor for one normal within the sampling set. On the other hand, it is even more important that the resulting discrete spherical function is independent towards the orientation of the model or the object stored in the normal map. Different sample sets are known that differently fulfill these demands and two of them are compared here. The *Clenshaw-Curtis Grid* and the *Fibonacci Spiral* on the sphere. The *Icosphere* is omitted here, as in [51] it was shown that it performs even worse than the equiangular approach.

The Clenshaw-Curtis Grid. When the sphere is sampled, using equispaced spherical coordinates, the result is the *Clenshaw-Curtis grid* $\mathcal{S}_{CC}$ [42]. The sample point coordinates $s_{i,j} \in \mathbb{S}^2$ can easily be computed by

$$s_{i,j} = (\varphi_i, \theta_j) = \left(\frac{i\pi}{S+1}, \frac{j\pi}{2S} \right), \text{ with } i \in \{0, \ldots, 2S\} \text{ and } j \in \{0, \ldots, 2S+1\}.$$

$$(5.62)$$

Here, $\varphi_i \in [0, 2\pi)$ is the azimuth angle and $\theta_j \in [0, \pi)$ is the polar angle. The integration weights $w_{i,j}$ that are needed to compensate the nonequispaced sample distances are defined as

$$w_{i,j} = \frac{4\pi\epsilon_j^{2S}}{S(2S+2)} \sum_{l=0}^{S} \epsilon_l^S \frac{1}{1-4l^2} \cos\left(\frac{jl\pi}{S} \right) \qquad (5.63)$$

with

$$\epsilon_j^J = \begin{cases} \frac{1}{2}, & \text{if } j = 0 \text{ or } j = J \\ 1, & \text{if } 0 < j < J. \end{cases} \qquad (5.64)$$

As the sample points are generated using equispaced azimuthal and polar angles, all points are located on equispaced circles on the sphere and the nearest neighbor of an arbitrary normal can be found in $\mathcal{O}(1)$. But, the resolution of the grid is dependent on the polar angle and the grid is not rotational invariant.

The Fibonacci Spiral. In contrast to the easy accessible but non-equispaced Clenshaw-Curtis grid, a very uniform sample set $\mathcal{S}_{FS}$ can be generated using the *Fibonacci spiral* [80]. The spiral can be built using an odd number of points $N = 2P + 1$. The point with the index i has the spherical coordinates

$$\varphi_i = 2\pi i \Phi^{-1},$$
$$\theta_i = \arcsin\left(\frac{2i}{N} \right) + \frac{\pi}{2}. \qquad (5.65)$$

In this equation $\Phi = \frac{1+\sqrt{5}}{2}$ is the golden ratio and $i \in \{-P, \ldots, P\}$. As the point distribution on the sphere can be assumed to be equispaced, the integration weights can be assumed to be constant and of the size of the surface patch around each point:

$$w_i = \frac{4\pi}{N}, i \in \{1, 2, \ldots, N\} \qquad (5.66)$$

Independent of N, only 10 areas near the north and south pole of the sphere differ by more than $\approx 2\%$ of this value [29]. As the distribution of points on the sphere is not as intuitive as the Clenshaw-Curtis grid, it is helpful to estimate the size of the surface patches defined through the point grid. As the area of each patch can be approximated as irregular hexagons, the inner angle θ of a triangle spanned by the center of the sphere and two points on the border of the hexagon can be approximated as

$$\theta = 4\sqrt{\frac{2\pi}{3\sqrt{3}N}}, \qquad (5.67)$$

cf. [32]. Therefore, 635 points on a Fibonacci sphere are necessary for an angular resolution of $10°$.

To overcome the binning complexity of the Fibonacci spiral, a *k-d-tree* can be used for normal binning. An alternative to the k-d-tree is the use of $\mathcal{S}_{CC}$. As the binning complexity for the Clenshaw-Curtis grid is $\mathcal{O}(1)$, it can be used as lookup table. In each point in $\mathcal{S}_{CC}$ a link to the nearest point in $\mathcal{S}_{FS}$ is stored. By this, the binning complexity of $\mathcal{S}_{FS}$ is also $\mathcal{O}(1)$ after an initial offline calculation. The amount S of grid points on $\mathcal{S}_{CC}$ has to be large enough to achieve an accurate lookup sampling. The biggest area of bins on $\mathcal{S}_{CC}$ is located at the equator with an area A_{max} of

$$
\begin{aligned}
A_{max} &= \int_{\frac{i\pi}{S+1}}^{\frac{(i+1)\pi}{S+1}} \int_{\frac{S\pi}{2S}}^{\frac{(S+1)\pi}{2S}} \sin(\theta) d\theta d\phi = \frac{\pi}{S+1} \left(\int_{\frac{\pi}{2}}^{\frac{\pi}{2}+\frac{\pi}{2S}} \sin(\theta) d\theta \right) \\
&= \frac{\pi}{S+1} \sin\left(\frac{\pi}{2S}\right).
\end{aligned}
\tag{5.68}
$$

For large S, it holds that $\sin\left(\frac{\pi}{2S}\right) = \frac{\pi}{2S}$. Therefore, to use $\mathcal{S}_{CC}$ as lookup table for an $\mathcal{S}_{FS}$ grid with N points, S has to be

$$
\begin{aligned}
\frac{4\pi}{N} &= M \frac{\pi}{S+1} \\
\Rightarrow S &= -\frac{1}{2} + \sqrt{\frac{1}{4} + \frac{\pi M N}{8}}.
\end{aligned}
\tag{5.69}
$$

Here, the factor M defines the ratio of how much smaller the largest cell of $\mathcal{S}_{CC}$ shall be in comparison to the average cell size in $\mathcal{S}_{FS}$. A visual comparison of the two sample sets can be seen in Fig. 5.3 and the analysis of rotational invariance can be seen in Fig. 5.4.

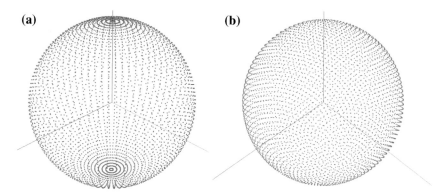

(a)   **(b)**

Fig. 5.3 Comparison of the $\mathbb{S}^2$ sample sets. **a** The Clenshaw-Curtis grid $\mathcal{S}_{CC}$. In the pole regions, clusters are visible. **b** The Fibonacci Spiral grid $\mathcal{S}_{FS}$, a nearly uniform sample set. Image taken with permission from [49]

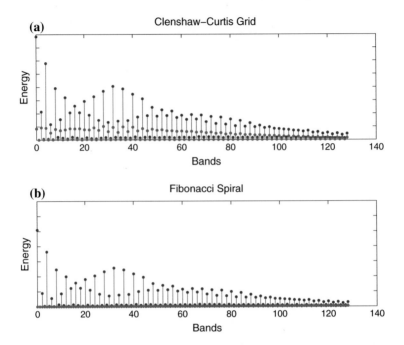

Fig. 5.4 Comparison of the energy of the EGIs using $\mathcal{S}_{CC}$ and $\mathcal{S}_{FS}$. *Blue* is the mean of the energy and *green* the standard deviation of the energy for each band. **a** Energy, using the Clenshaw-Curtis grid $\mathcal{S}_{CC}$. **b** Energy, using the Fibonacci Spiral grid $\mathcal{S}_{FS}$. The rotational variance of $\mathcal{S}_{CC}$ is obvious

Sample Sets for $SO(3)$. The result of the correlation of the two EGIs is element of $SO(3)$. Therefore, it is important to also sample $SO(3)$ uniformly. Moreover, it is important for the grid to build small neighborhoods, i.e. to not build clusters, to be able to perform an efficient maximum search. In other words, a good sample set builds a sample distribution in which each sample has as few neighbors as possible in a local neighborhood. One measure for the uniformity of a sample set $\mathcal{R}$ is its dispersion

$$D(\mathcal{R}) := \max_{r \in SO(3)} \min_{s \in \mathcal{R}} d(r, s), \tag{5.70}$$

where $d(r, s)$ is a metric on $SO(3)$ like described in Eq. (5.15).

Similar to the $\mathbb{S}^2$ sample sets, the straight forward way of sampling $SO(3)$ is an equiangular grid $\mathcal{R}_E$, using uniformly distributed Euler angles:

$$R\left(\alpha_i, \beta_j, \gamma_k\right) = R\left(\frac{2\pi i}{N}, \frac{\pi j}{N}, \frac{2\pi k}{N}\right); i, j, k \in \{1, \dots, N\} \tag{5.71}$$

Unfortunately, like $\mathcal{S}_{CC}$, $\mathcal{R}_E$ builds clusters.

It is therefore better, to construct a uniform sampling in $SO(3)$. This can be done by using the subgroup algorithm described in [19, 59]. In short terms, the subgroup algorithm divides $SO(3)$ into uniform partitions, each sampled uniformly. The partitions are the subgroup $\mathcal{R}_Z := \{\boldsymbol{R}_Z(\theta)|\theta \in [0, 2\pi)\}$ of rotations around the z-axis and the set $\mathcal{R} := \{\boldsymbol{R}_Z(\alpha)\boldsymbol{R}_Y(\beta)|\alpha \in [0, 2\pi), \beta \in [0, \pi)\}$ which can be interpreted as spherical coordinates. To achieve a near uniform sample set $\mathcal{R}_S$ of N points on $SO(3)$, R points have to be distributed on $\mathcal{R}_Z$ and S points have to be distributed on $\mathcal{R}$, so that $N = RS$. For the angular resolutions $\psi_{\mathcal{R}_Z}$ and $\psi_{\mathcal{R}}$, S and R have to be chosen as

$$\psi_{\mathcal{R}_Z} = \sqrt{\frac{2\pi}{S}},$$
$$\psi_{\mathcal{R}} = \frac{4\pi}{R}. \tag{5.72}$$

For $\psi = \psi_{\mathcal{R}_Z} = \psi_{\mathcal{R}}$ follows that

$$S = \sqrt[3]{N\pi},$$
$$R = \frac{N}{\sqrt[3]{N\pi}}. \tag{5.73}$$

Figures 5.5 and 5.6 show a comparison of the two sampling schemes. It can be seen that the subgroup algorithm based sampling does not build clusters of samples in $SO(3)$, in contrast to the equispaced Euler angle distribution.

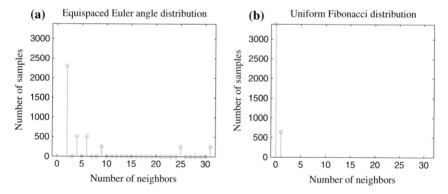

Fig. 5.5 $SO(3)$ local neighborhood comparison. Number of neighbors in a local neighborhood of $15°$ for all sample points. **a** Equiangular sampling with $N = 16$ (Eq. (5.71)), therefore 4096 sample points. **b** Subgroup algorithm based sampling using the Fibonacci spiral as sphere distribution. 4025 sample points overall, with 175 points on $\mathcal{R}$ and 23 equiangular distributed angles on $\mathcal{R}_Z$

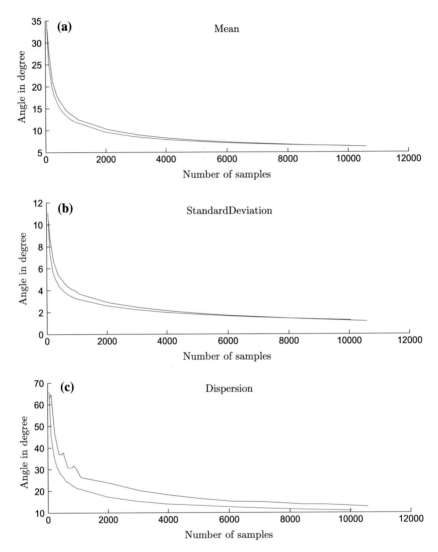

Fig. 5.6 $SO(3)$ sampling comparison. The distance between two sample points in $SO(3)$ is shown. $\mathcal{R}_E$ is plotted in *blue*, $\mathcal{R}_S$ is plotted in *red*. **a** Mean angle between two samples. **b** Standard deviation of angles between two samples. **c** Dispersion of the sampling

Fig. 5.7 Scan of an object out of one direction. The surface patches with normals pointing towards the camera are well scanned, the rest is invisible

Error Sources

Handling normal maps generated by one camera, which include faces with normals pointing directly to the camera can be well measured; whereas faces with growing angles between their normals and the camera's viewing direction often cannot be measured due to self occlusion. This may lead to a situation in which one normal is dominant in the EGIs, with the result that only this dominant normal can be oriented correctly and one degree of freedom, namely the orientation around this normal, stays unknown.

Figure 5.7 shows a scan using the same setup as the normal map acquisition setup but a CLA scan. It is obvious that one surface normal will be dominant in the according EGI. Fortunately, the resulting orientation error can be eliminated during the translation estimation step, which will be described in the following section.

5.2.2 Accurate Monocular Translation Estimation

With known orientation R of the model w.r.t. the camera, the translation of the model relative to the camera's coordinate system has to be estimated. As the model contains 3D data and the image 2D information, either the model data has to be reduced to 2D or the image data has to be expanded to 3D to get a comparable data base. Both of these techniques are possible and will be presented in the following sections.

2D Model Data and 2D Image Data

When the orientation R of the model w.r.t. the camera is known, the appearance of the object in the camera image can be approximated by a rendering of the object. As a perspective camera is used and the position of the object relative to the camera is unknown, an orthographic projection of the model onto the image plane is done to generate a single comparable representation of the model. The parallel projection only approximates the real appearance, but is similar enough to be used for the translation estimation, even in presence of possible rotation estimation errors, as described in Sect. 5.2.1.

The two images to be compared can be regarded as functions $i_n : \mathbb{R}^2 \to \mathbb{S}$ and $i_o : \mathbb{R}^2 \to \mathbb{S}$ representing the normal image and the rendered image of the model, respectively. Now the translation of the model can be estimated via scale invariant correlation of the rendering and the acquired normal map.

Ignoring perspective distortions, the relation of i_n and i_o can be written as

$$i_o(x) = i_n(Dx + t) \tag{5.74}$$

where $x = [x, y]^T$ are the image coordinates, $t = [t_x, t_y]^T$ a 2D translation in the image plane and $D = sR$ and $R = \begin{bmatrix} \cos(\omega) & \sin(\omega) \\ -\sin(\omega) & \cos(\omega) \end{bmatrix}$ is a rotation and scaling matrix with rotation angle ω and scaling factor s.

Image Based Scale Estimation with Orientation Correction. Equation (5.74) transformed into the Fourier domain leads to

$$I_o(k) = \frac{1}{|det(D)|} I_n\left(\frac{1}{s}Rk\right) e^{j\frac{1}{s}(Rk)^T t} \tag{5.75}$$

where I_o is the spectrum of i_o, $I_o = \mathcal{F}\{i_o\}$, I_n is the spectrum of i_n, $I_n = \mathcal{F}\{i_n\}$, multi-index $k = (k_x, k_y)^T$, and $\mathcal{F}\{\cdot\}$ denotes the Fourier transform. The determinant of D is s^2, as $det(R) = 1$. It can be seen that the translational offset t results in a phase shift $e^{j\frac{1}{s}(Rk)^T t}$ in the spectrum I_n. To break down the problem and solve it for the individual unknowns (t_x, t_y, s, ω), the magnitude spectra $M_o = |I_o|$ and $M_n = |I_n|$ are used. As the translation offset in the image corresponds to a phase shift in the spectrum (translation property), the use of only the magnitudes will eliminate the translational parameter.

$$M_o(k) = \frac{1}{s^2} M_n\left(\frac{1}{s}Rk\right) \tag{5.76}$$

The two unknown parameters ω and s are present as rotation and scaling in this equation. By a coordinate transform of the spectra into polar coordinates

$$(\varphi, r)^T = \left(\tan^{-1}\left(\frac{k_y}{k_x}\right), \sqrt{k_x^2 + k_y^2}\right)^T \tag{5.77}$$

the rotation around ω transforms to a shift along the φ-axis. To further transform the scaling along the r-axis to a shift, the logarithm of the radii is used resulting in a logarithmic-polar representation of the spectra [83].

$$M_o^{lp}(k) = \frac{1}{s^2} M_n^{lp}\left(k - [\log(s), \omega]^T\right) \tag{5.78}$$

To estimate the relative translation of M_o^{lp} and M_n^{lp}, which is simply a shift of the images, a 2D correlation of both signals is computed.[3]
Correlation of Functions on $\mathbb{R}^2$. The relation between the images in Eq. (5.78) is a simple shift. A cross correlation of the two functions can solve for the present

---

[3]As the orientation is known at this point, the parameter ω equals 0 as this parameter describes the rotation of the model in the image plane. Nevertheless, with ω a possibly erroneous DOF, as

translational offset. For sake of simplicity and generality renaming $M_o^{lp}(\mathbf{k})$ as $i_1(\mathbf{x})$ and $M_n^{lp}(\mathbf{k})$ as $i_2(\mathbf{x})$, the relation of the two images can be written as

$$i_2(\mathbf{x}) = i_1(\mathbf{x} + \mathbf{t}).\tag{5.79}$$

To solve for $\mathbf{t}$, the cross correlation of the two functions can be computed, defined as

$$c(\boldsymbol{\xi}) = i_1(\boldsymbol{\xi}) * i_2(-\boldsymbol{\xi}) = \int_{\mathbb{R}^2} i_1(\mathbf{x}) i_2(\mathbf{x} + \boldsymbol{\xi}) d\mathbf{x}.\tag{5.80}$$

Using the convolution theorem of the FFT and with the Fourier transform $\mathcal{F}\{\cdot\}$, the inverse Fourier transform $\mathcal{F}^{-1}\{\cdot\}$, and $I_1 = \mathcal{F}\{i_1\}$ and $I_2 = \mathcal{F}\{i_2\}$, the cross correlation can alternatively be obtained by

$$c(\boldsymbol{\xi}) = \mathcal{F}^{-1}\left\{ I_1(\mathbf{k})\overline{I_2(\mathbf{k})} \right\}.\tag{5.81}$$

Here, $\overline{I(\mathbf{k})}$ denotes the conjugate complex of $I(\mathbf{k})$.

More robust techniques to image correlation are widely available in literature, such as orientation correlation [24], phase correlation [3] and gradient cross-correlation [4]. The latter is used for the computations in this thesis, and a short summary is given in the following paragraphs.

The gradient correlation (GC) is based on image gradients and takes their orientation as well as their magnitude into account. A gradient image, in this context, can be seen as complex image $g : \mathbb{R}^2 \to \mathbb{C}$.

$$gi(\mathbf{x}) = g_{i,x}(\mathbf{x}) + j g_{i,y}(\mathbf{x})\tag{5.82}$$

$g_{i,x} = \nabla_x i_i$ and $g_{i,y} = \nabla_y i_i$ are the partial derivatives of the corresponding image i_i. With this, the GC can be written as

$$gc(\boldsymbol{\xi}) = g_1(\boldsymbol{\xi}) * \overline{g_2(-\boldsymbol{\xi})} = \int_{\mathbb{R}^2} g_1(\mathbf{x})\overline{g_2(\mathbf{x} + \boldsymbol{\xi})} d\mathbf{x}.\tag{5.83}$$

In the Fourier domain, the problem becomes

$$GC(\mathbf{k}) = \mathcal{F}^{-1}\left\{ G_1(\mathbf{k})\overline{G_2(\mathbf{k})} \right\}.\tag{5.84}$$

(Footnote 3 continued)
described above, can be estimated. The rotation frame that has been estimated prior has to be updated with ω, if $\omega \neq 0$. This update has to be a rotation around the dominant peak of s_n.

This correlation, performed as multiplication in the Fourier domain gives a very accurate estimation of the relative position of i_2 relative to i_1. In the context of scale and rotation error estimation, the maximum of $gc(x)$ results in an estimate for $\left[\log(s), \omega\right]^T$.

Image Based Translation Estimation. At this point, the scaling of the object is known and a possible orientation estimation error has been eliminated. Thus, with the known parameters ω and s a scaled and rotated image i'_o can be computed as

$$i'_o(x) = i_o \left(sx \begin{bmatrix} \cos(\omega) & \sin(\omega) \\ -\sin(\omega) & \cos(\omega) \end{bmatrix} \right). \tag{5.85}$$

So the relation of the model image and the normal image now simplifies to

$$i'_o(x) = i_n(x + t). \tag{5.86}$$

The translation vector t can now be estimated using gradient correlation as already described above. The maximum of $gc(x)$ in this context gives the position of the scaled model image relative to the original normal map. Using the camera calibration, the resulting pixel coordinate t defines a 3D viewing ray v, along which the object is translated away from the focal point of the camera. The distance of this translation can be computed via the intercept theorem and scaling factor s (Fig. 5.8).

$$\frac{d_o}{d_n} = \frac{l_o}{l_n} = s \tag{5.87}$$

The translation of the object to the focal point results in $t_{\text{pose}} = d_o v$.

  Here, d_o is the distance of the object to the focal point, d_n the distance of the pixel to the focal point and l_o and l_n are the real size of the object and the size of the object in the normal map respectively.

  The values of the position estimation correlation maxima serve as a quality estimate for the estimated pose.

  When a position is found, it is possible to render the model using the approximately correct perspective. With the new rendering, the search procedure can be repeated to acquire a better accuracy as the projection and the image are more similar.

Fig. 5.8 Computation of the distance d_o of the object to the focal point f of the camera using the intercept theorem

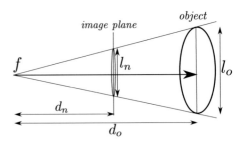

Advantages and Problems. The approach described above is based on 2D images only. This means that all computations can be computed using 2D images and the Fast 2D Fourier transform which makes the computation efficient. The solution is analytically tractable and has a predictable run-time. Furthermore, a fourth parameter, additional to the three translation parameters, can be estimated for a possible error correction.

But, there are drawbacks to this approach. The main problem is that the approach does not offer a closed solution for the three translational parameters. The scale estimation is decoupled from the viewing ray estimation but both are needed for the three Cartesian degrees of freedom of the translational part of the pose of the object. In the case that the combination of the global maximum of the scale estimation and the global maximum of the viewing ray estimation does not deliver a satisfying translation estimation, another optimal combination of the two correlation functions has to be found. As the viewing ray estimation is based on the scale estimation for each local scale maximum $s_{\max,i}$, a single viewing ray estimation has to be computed. Another problem is that without position estimation, which is the purpose of the computations, a perspective projection of the object is impossible. Therefore, the projected image only approximates the appearance of the model in the normal image, which alters the quality of the translation estimation prior to the refinement step.

3D Model Data and 3D Image Data

When reducing the dimensionality of the model data from 3D to 2D, information gets lost. Even worse, the projection to the image plane results in perspective errors that reduce the quality of the registration, as mentioned above. Therefore, the dimensionality reduction should be avoided. This means that the data of the monocular camera have to be expanded from 2D to 3D. Regarding the camera model, when 3D information is needed, a camera does deliver 3D information. More precisely, the camera image defines an array of 3D viewing rays. A complete frustum can therefore be built with constant normals along each viewing ray. This frustum is stored as a 3D voxel grid. Each voxel contains the normal of the pixel that defines the viewing ray intersecting the voxel, see Fig. 5.9.

Additionally the mesh of the model is converted into a voxel grid. Using the same voxel size as before, all voxels are assigned to the normals of the triangles intersecting these voxels, see Fig. 5.10. As each viewing ray in the frustum of the camera points to a surface with a surface normal pointing towards the camera, only normals with an angle $\angle(\boldsymbol{n}_{o,i}, \boldsymbol{z}_c) < \frac{\pi}{2}$ are stored in the grid. $\boldsymbol{n}_{o,i}$ is the specific surface normal and $\boldsymbol{z}_c$ the z-axis of the camera. Two volumes $v_i : \mathbb{R}^3 \to \mathbb{S}$ and $v_o : \mathbb{R}^3 \to \mathbb{S}$ can be obtained by this procedure. Through the construction of the viewing frustum, the perspective projection character of the camera is converted to a regular Cartesian voxel grid. The voxel representation of the model is regular Cartesian, too, which makes the two representations perfectly comparable.

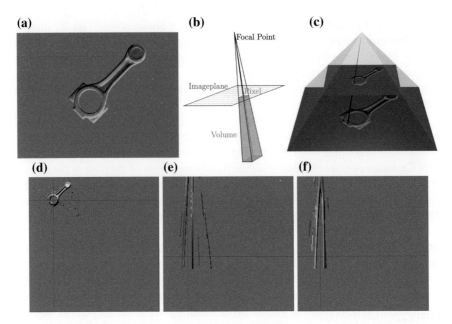

Fig. 5.9 Frustum of a normal map, zero-padded to match the volume sizes of model and frustum. Only the azimuthal angle of the surface normals is visualized. **a** Normal image acquired using photometric stereo. **b** Schematic plot of the volume conversion for a single pixel. **c** 3D frustum of the normal image built using the camera calibration parameters. **d** x-y-plane slice of the frustum. **e** x-z-plane slice of the frustum. **f** y-z-plane slice of the frustum

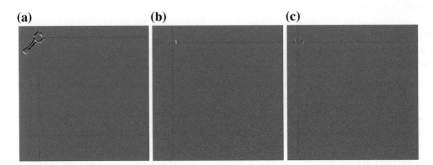

Fig. 5.10 Voxelized model, zero-padded to match the volume sizes of model and frustum. The volume is mirrored as preparation for the convolution in the Fourier domain. Only the azimuthal angle of the surface normals is visualized. **a** x-y-plane slice of the volume. **b** x-z-plane slice of the volume. **c** y-z-plane slice of the volume

The relation between the two volumes can be written as simple translation

$$v_o(\boldsymbol{x}) = v_i(\boldsymbol{x} + \boldsymbol{t}) \cdot \chi(\boldsymbol{x}), \text{ with } \chi(\boldsymbol{x}) = \begin{cases} 1, & \text{if } v_o(\boldsymbol{x}) \neq \boldsymbol{0} \\ 0, & \text{otherwise} \end{cases}. \tag{5.88}$$

Data: correlation volume V_c
Result: global maximum $v_{max, refine}$
find global minimum M_v in V_c;
$m_z \leftarrow M_v.z$;
for $z = m_z - range_z \ldots m_z + range_z$ **do**
$\quad$ find global maximum $M_{p,z}$ in z-plane of V_c;
$\quad$ calculate weighted mean $\tilde{M}_{p,z}$ in 2D local neighborhood around $M_{p,z}$;
end
calculate weighted mean $v_{max, refine}$ of all $\tilde{M}_{p,z}$;

Algorithm 1: Sub-voxel accurate maximum estimation.

The solution to this problem is exactly the same as in the two dimensional case. The maximum of the cross-correlation of the two volumes delivers a translation estimation of the model relative to the viewing frustum, i.e. the camera. The computation of the correlation can also be efficiently computed in the Fourier domain.

$$c(\boldsymbol{\xi}) = \mathcal{F}^{-1}\left\{V_i(\boldsymbol{k})\overline{V_o(\boldsymbol{k})}\right\} \tag{5.89}$$

As the correlation function, in contrast to the 2D-2D variant, is present as single discrete function on $\mathbb{R}^3$, the local surroundings of the global maximum can be analyzed to achieve sub-voxel accuracy of the translation estimation result. This can, for example, be done by a weighted mean of the values in a local neighborhood around the global maximum. As the inaccuracy, which can be seen in Fig. 5.11, is mainly present along viewing rays, the neighborhood is defined along these rays. See Algorithm 1 for details.

Advantages and Problems. The main advantage of this approach is that the translation estimation is contained in a closed form solution. All three parameters are computed at once and, therefore, a complete solution space is available. Within the

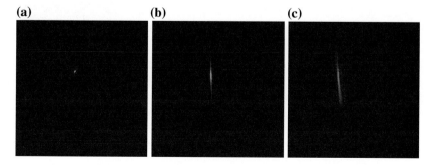

(a)   (b)   (c)

Fig. 5.11 Correlation function of the two volumes of Figs. 5.9 and 5.10. The volume is centered at the global maximum. It can be seen that the maximum is much sharper in the x-y-plane as in z-direction. The neighborhood analysis takes the complete maximum structure into account. **a** x-y-plane slice of the volume. **b** x-z-plane slice of the volume. **c** y-z-plane slice of the volume

solution space, a global maximum can be found with sub-voxel accuracy, considering an arbitrarily shaped neighborhood. This methodology is capable of overcoming the main systematic problem of monocular distance estimation. This systematic problem can be described as: "*Given a minor change of the object's pose in viewing direction of the monocular camera, no change of the image can reliably be captured.*" But, if a larger neighborhood of the correct position is known, a larger translation of the object along the viewing direction can be analyzed and a global optimum can be estimated using larger differences in the camera image.

The only disadvantage of this approach is the amount of data which needs to be held in the memory. As the volumes contain complex values, the resolution of the volumes has to be held quite small to keep computation time and memory utilization within reasonable limits.

5.3 Bin-Picking Application—Collision Avoidance

The biggest issue of using normal maps for autonomous robot manipulation is that no 3D data of the scene might be available. This makes the collision avoidance procedure quite complicated. In a structured scenario, like industrial bin-picking, where obstacles in the workspace of the robot are mainly known, a collision avoidance can be performed, nevertheless.

All known obstacles have to be 3D modeled and considered in the collision avoidance. By this, the only unknown obstacles are the parts to be gripped, i.e. the objects in the bin or on the table, etc. When all possible objects are known by their models, the pose estimation technique proposed in this chapter can be used to locate all objects in the scene.[4] When all objects are located, their models can be transformed to their specific world poses and a 3D collision avoidance can be performed, using the same approaches as presented in Sect. 3.2, including the grasp planning technique.

5.4 Experimental Normal Map Based Grasping

Like the two pose estimation techniques described in the former chapters, normal maps were used to locate and autonomously grasp objects. As the photometric stereo method was used, all objects in the real world scenarios were painted matte white to generate known reflectance properties. A detailed look at the single-shot multi-spectral photometric stereo is given in the Appendix A.2. Furthermore, only isolated objects were grasped. Prior to the robotic grasping application, a set of simulated scenes were used to analyze the possible accuracies of the system.

---

[4]The experiments will show that the approach is capable of locating several objects.

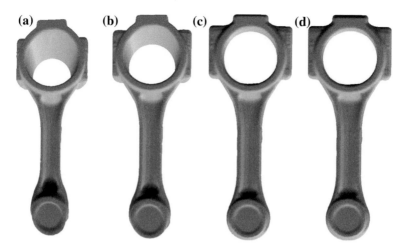

Fig. 5.12 Impact of the focal length on the appearance of a model (piston rod) in the image.
a $f = 1$ mm, distance = 75 mm **b** $f = 4$ mm, distance = 300 mm **c** $f = 12.5$ mm, distance = 900 mm
d $f = 32$ mm, distance = 2400 mm

5.4.1 Simulation

As the setup in the robotic work cell contained an industrial camera, the same camera
was simulated in a virtual environment. The most important property of the camera
is the focal length of the attached lens. Very short focal lengths significantly distort
the scene whereas long focal lengths result in very low perspective scaling. As the
perspective is essential for rotation (weak perspective assumption) and translation
estimation (perspective scaling) the analysis of the impact of the focal length on
the localization result is important. The experiments were therefore performed using
focal lengths between 1 and 64 mm.[5] Note that the 1/3 inch sensor of the camera has
a crop factor of 8, which means that the 35 mm equivalent focal lengths are between
a fish-eye (8 mm) and a super telephoto lens (512 mm). At the short end the models
are significantly distorted but the perspective scaling is large. At the long end the
viewing rays are nearly parallel, which results in nearly zero perspective scaling and
distortion. All scenes were rendered such that the objects approximately filled 10 %
of the frame. As the focal length obviously changes the field of view, the distance
between the camera and the object varied from 75 at 1 mm focal length to 4800 at
64 mm focal length (see Fig. 5.12).

Orientation Estimation Accuracy

The orientation estimation is the first of the two steps performed for pose estimation. Based on the result, the translation is estimated. Therefore, it is essential that
the orientation estimation is robust and accurate. To analyze the accuracy, different

---

[5]The simulated focal lengths were 1, 2, 4, 8, 12.5, 16, 32 and 64 mm.

models and different parameters were analyzed for their impact on its results. The most important aspect is the focal length of the camera. As the focal length defines the possible working distance, it has to match the task and may not be chosen freely. But, for short focal lengths, the weak perspective assumption does not hold. Furthermore, the robustness against noise is important. Both aspects have been analyzed. At first, the normalized cross correlation was used. The experiments showed that the best parameters in the current implementation were achieved, using a bandwidth of $L = 8$, an amount of points on the EGIs of $|\mathcal{S}_{FS}| = 309$, and an amount of points in $SO(3)$ of $|\mathcal{S}_{S}| = 9957$. Using this small amount of points on $\mathbb{S}^2$ results in an inner angle between two neighboring sample points of $14.3°$, but also a very short computation time of only 0.5 s. The results were produced using 51 simulated normal maps generated using arbitrary camera poses. The objects were placed in the field of view, so that they filled approx. 10% of the image. To compare the rotations, the metric defined in equation (5.15) was used. A result was defined to be successful if one of the first 20 maxima in the correlation differed at most $25°$ from the ground truth.[6] The experiments show that the results are nearly independent of the used focal length (see Fig. 5.13). Only when very short lenses are used, the results get worse.

Considering the noise level, it can be seen that noise slightly effects the results (see Fig. 5.14). The applied noise is normally distributed and given by its standard deviation. It is noticeable that the orientation estimation gets better when a small amount of noise is present in the normal maps. This is due to the nearest neighbor binning. The normals are only stored in the nearest bin. Neighboring bins are not considered. Therefore, slight rotations of the object in the image can effect the appearance of the according EGI. This effect is reduced by the added noise.

Overall, it can be stated that the algorithm performs very robustly. Only very complex objects need the analysis of many local correlation maxima to find a correct orientation. To overcome this issue, the alternative of the energy lookup proposed in Sect. 5.2.1 was analyzed (see Fig. 5.15). For the tests, the same setup as above was used (bandwidth $L = 8$ to create the energy feature vectors, $|\mathcal{S}_{FS}| = 309$, $|\mathcal{S}_{S}| = 9957$). The lookup table contained 127 EGIs. The generation of the lookup was computed in 70 s, using un-optimized code. The lookup table can be computed offline and does not alter the overall run time. After the search for a nearest neighbor in the lookup table, the same correlation as before is computed to get an orientation estimate.

The performance of the approach, using simple objects, is much worse than before. The problem is that the energy vector is invariant w.r.t. symmetries. The cube, the piston rod, as well as the metal part are subject to symmetries. The lookup for these objects may therefore result in erroneous viewpoints and wrong EGIs. These problems could be avoided by a more sophisticated search strategy. When a nearest neighbor is found, the mirrored entry of that match may be as good as the entry itself. Therefore, both entries should be considered and the correct one could then be chosen by the maximum value of the subsequent correlation.

---

[6]With a rotation error of at most $25°$ the appearance of the model in the image is still quite similar to the acquired image.

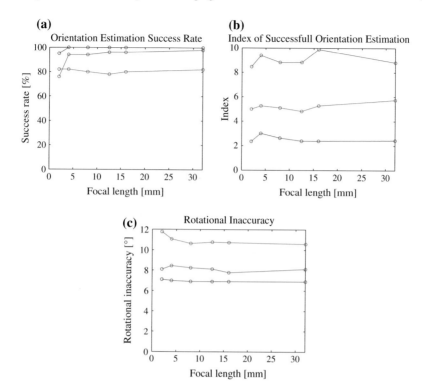

Fig. 5.13 Analysis of the accuracy of the orientation estimation technique dependent on the focal length of the camera for different objects of the industrial parts data set. The results for the piston rod are *blue*, the results for the metal part are *red*, and the results for the balance shaft are *green*. **a** Success rate of the estimation. A success is a result that is less than 25° away from the ground truth and within the first 20 maxima in the correlation. **b** Average index of the correlation maximum, describing a successful orientation estimation, considering ordered maxima in the correlation function. **c** Average rotation error of the successful orientation estimation attempts

Complex objects, by contrast, like the dragon model, perform very well using this technique, as the hidden parts in the model's EGI are omitted and the similarity of both EGIs is much higher.

Overall, the robustness against noise is much worse than before. This is because noisy normals have a great effect on the energy vector. Here, and also in the normalized cross correlation, it can be sensible to smooth the EGIs, prior to the lookup and the correlation. In [43] a technique is proposed that could be applied.

Translation Estimation Accuracy

The accuracies of the image based translation estimation for three different objects using the "2D-2D method" can be seen in Figs. 5.16, 5.17 and 5.18. These objects have a very complex shape (Armadillo), a quite spherical shape (Bunny), or a longish shape (piston rod) to cover different types of geometries. The results were produced

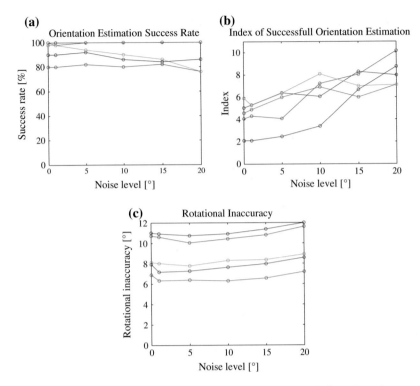

Fig. 5.14 Analysis of the accuracy of the orientation estimation technique dependent on the noise level present in the normal map, using a 16mm lens. The results for the cube are *blue*, the results for the piston rod are *green*, the results for the dragon are *purple*, the results for the balance shaft are *red*, and the results for the metal part are *cyan*. **a** Success rate of the estimation. A success is a result less than 25° away from the ground truth and within the first 20 maxima in the correlation. **b** Average index of the correlation maximum, describing a successful orientation estimation, considering ordered maxima in the correlation function. **c** Average rotation error of the successful orientation estimation attempts

using 100 translation estimation procedures per focal length with given orientation estimation. The orientation used for translation estimation was distorted by orientations of 1° in each axis to generate more realistic results.

All models caused very similar results matching intuitive interpretations. At very short focal lengths, the distortion of the object in the image is so high that the similarity between the parallel projection used for translation estimation and actual image is not high enough. This results in high translation errors in all three directions. At very long focal lengths, the perspective scaling of the model in the camera image gets very low. Therefore, small translations along the viewing direction (z-axis) cause only marginal scalings of the model in the image. The results show that the translation error is mainly in z-direction in these cases. Furthermore, in both of the extreme cases, the perspective correction does not notably refine the results. In case

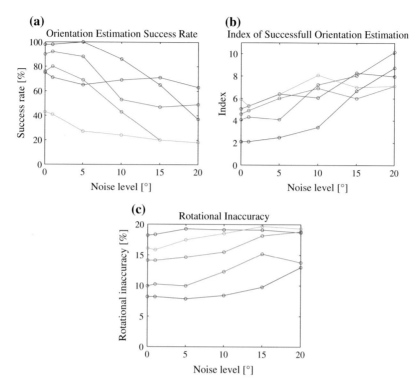

Fig. 5.15 Analysis of the accuracy of the orientation estimation technique, based on an energy lookup, dependent on the noise level present in the normal map, using a 16 mm lens. The results for the cube are *blue*, the results for the piston rod are *green*, the results for the dragon are *purple*, the results for the balance shaft are *red*, and the results for the metal part are *cyan*. **a** Success rate of the estimation. A success is a result less than 25° away from the ground truth and within the 20 first best results. **b** Average index of the correlation maximum, describing a successful orientation estimation, considering ordered maxima in the correlation function. **c** Average rotation error

of short focal lengths, this is due to the extreme distortions. When the first translation estimation is not accurate enough, the perspective projection of the model onto the image plane is not better than a parallel projection. In case of long focal lengths, the perspective projection is very similar to the parallel projection and therefore does not refine the result.

Nevertheless, the experiments show robust results for moderate focal lengths and better results than expected even at long distances. With an average accuracy of under 25 at 4800 mm distance between camera and model, the accuracy of the estimation approach is still comparable to modern laser scanner accuracies.[7]

---

[7]The SICK LMS 500 laserscanner [76] has a systematical error of ±25 mm and a statistical error of ± 7 mm in the range of 1 m, ..., 10 m.

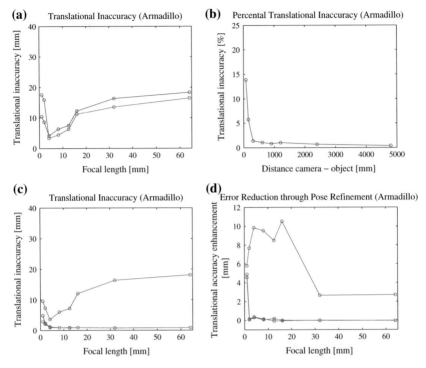

Fig. 5.16 Analysis of the position accuracy of the image based translation estimation technique dependent on the focal length of the camera and an orientation error of $1°$ in each axis for the Armadillo model. **a** Absolute overall translation error (mean error in *blue*, standard deviation in *green*). **b** Relative overall translation error dependent of the distance between camera and object. **c** Absolute errors in each axis (*blue* x-axis, *green* y-axis, *red* z-axis). **d** Error reduction achieved, after a second translation estimation using a perspective rendering of the model using the first translation estimation result

Besides the focal length, the translation estimation accuracy is dependent on the quality of the rotation estimation. This dependency was analyzed for different models. In Fig. 5.19, the results are shown for the armadillo, the bunny, and the piston rod model. It can be seen that the errors rise linearly with the rotation error. Furthermore, it can be seen that the quality depends on the model. This is due to the general shape. For near spherical objects, like the bunny, the size of the model in the image is quasi independent of the rotation. For the piston rod, the difference is bigger and therefore the translation error, too.

In comparison to the image based translation estimation, the volume based ("3D-3D") translation estimation technique was also analyzed, using the same technique as above. The results can be seen in Fig. 5.20. The experiments show that the volume based technique is neither more accurate nor more reliable than the image based technique. This is true for the actual implementation of the technique. In contrast

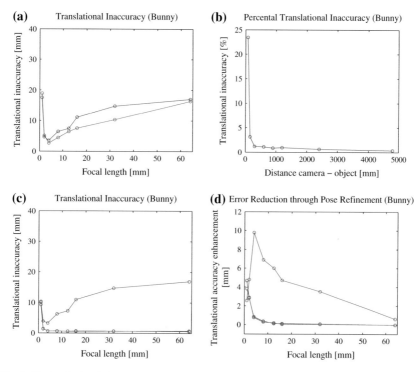

Fig. 5.17 Analysis of the position accuracy of the image based translation estimation technique dependent on the focal length of the camera and an orientation error of 1° in each axis for the Bunny model. **a** Absolute overall translation error (mean error in *blue*, standard deviation in *green*). **b** Relative overall translation error dependent of the distance between camera and object. **c** Absolute errors in each axis (*blue* x-axis, *green* y-axis, *red* z-axis). **d** Error reduction achieved, after a second translation estimation using a perspective rendering of the model using the first translation estimation result

to the image based technique, the resolution of the volumes is by far lower than the resolution of the images. This is due to the higher memory consumption of the three dimensional volumes.

5.4.2 Real World Scenario

To test the real world applicability of the normal map based pose estimation technique, the already introduced industrial setup was used, again. As the localization technique was only tested for isolated objects, only single piston rods were grasped in the experiments. To generate the normal maps, an enhanced multi spectral photometric stereo method was used. The method is briefly described in the appendix in Sect. A.2.2. The industrial camera "Imaging Source DFK 31BF03" with a resolution

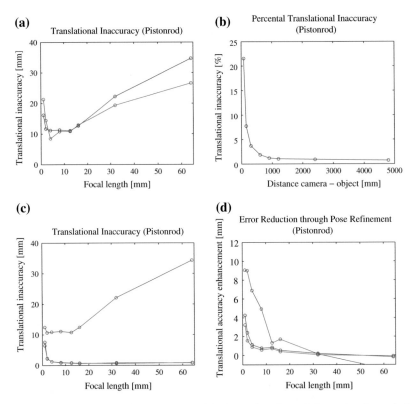

Fig. 5.18 Analysis of the position accuracy of the image based translation estimation technique dependent on the focal length of the camera and an orientation error of 1° in each axis for the Pistonrod model. **a** Absolute overall translation error (mean error in *blue*, standard deviation in *green*). **b** Relative overall translation error dependent of the distance between camera and object. **c** Absolute errors in each axis (*blue* x-axis, *green* y-axis, *red* z-axis). **d** Error reduction achieved, after a second translation estimation using a perspective rendering of the model using the first translation estimation result

of 1024 × 768 pixels was used, equipped with a 16 mm lens. The focal length of the lens resulted in a working distance of 120 cm between the camera and the objects. To simplify the data acquisition, the piston rods were painted white to generate known reflection properties. The same paint was then used on a calibration body to fill the lookup table of the photometric stereo camera. The normals in the normal maps had an average error of 12°. The acquisition time was equal to the exposure time of the camera.

The work cell setup can be seen in Fig. 5.21. As only single objects were grasped, no real bin-picking was performed and no collision avoidance mechanisms were included into the system. To enhance the system for bin-picking, a segmentation algorithm has to be included. In a series of 50 grasp attempts, a success rate of 94 %

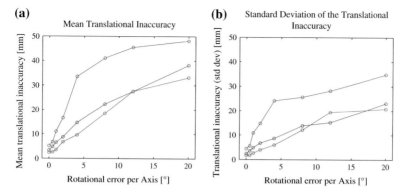

Fig. 5.19 Dependency of the image based translation estimation result of the rotation estimation quality. The given rotation error is an error in all three axes. The *blue bars* are for the Armadillo model and focal length 8 mm, the *green bars* for the bunny model and focal length 4 mm and the *red bars* for the piston rod model and focal length 12.5 mm. **a** Mean overall translation error. **b** Standard deviation of the overall translation error

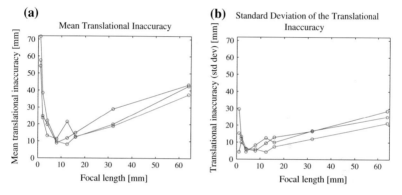

Fig. 5.20 Dependency of the volume based translation estimation result of the focal length. The *blue bars* are for the Armadillo model, the *green bars* for the bunny model and the *red bars* for the piston rod model. **a** Mean overall translation error. **b** Standard deviation of the overall translation error

was achieved. The three failed attempts were caused by an inaccurate z-coordinate of the object pose. Examples of the test scenarios can be seen in Fig. 5.22. As 10 maxima of the orientation estimation correlation were analyzed, each with 2 maxima in the scale estimation correlation, the overall pose estimation time was 50 s. Note that no abort criterion was used to enhance this time to get the best possible pose estimation.

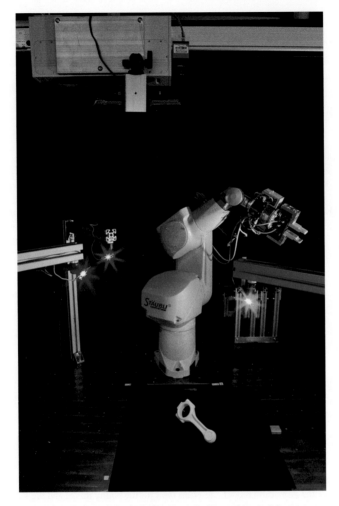

Fig. 5.21 Robot work cell equipped with a photometric stereo based vision sensor

5.5   Discussion

This chapter describes an approach to localize objects in 3D using normal maps. In
the most extreme case, a single camera and a set of at least three LED lights is enough
for 3D object localization using this technique. Moreover, in scenarios in which only
normal maps are available, 3D object localization can now be performed, which opens
possibilities for new types of sensors to be used for pose estimation. For example, in
critical scenarios in which laser scanners cannot be used for safety reasons, like house
hold applications, this can be a big advantage. As a two step approach is presented,
both steps of the algorithm can be applied without each other. When for example the

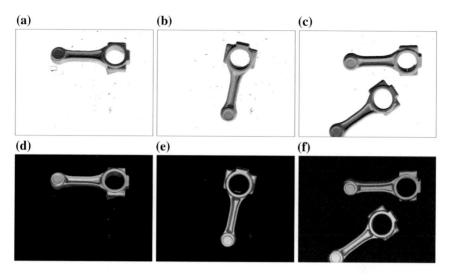

Fig. 5.22 Results of the normal map based pose estimation technique on real data. The results are superposed by a *red* rendering of the model on the *green* normal map. As can be seen, even in presence of more than one object, correct object poses are determined. **a, b** Normal map of a scene with isolated object. **c** Normal map of a scene with two objects. **d–f** Localization results

orientation of objects in a camera image are known, the 3D translation estimation procedure can be used for accurate monocular position estimation.

The normal map approach cannot compete with approaches based on 3D data w.r.t. accuracies. Nevertheless, considering the fact that a 6D pose is computed using only the derivatives of 3D surfaces, the accuracies are quite high, i.e., they are high enough to be used for automation tasks. The main problem is the long run time of the algorithms. There are many possibilities to optimize the run time but these were not scope of this thesis.

When it comes to physical interaction with the located objects, another problem is that no 3D scan data is available for collision avoidance. This problem can only be tackled by known environment models and a localization of all objects within the environment or by additional sensors in the gripper of the robot.

The algorithms were only tested on quite simple scenes. Further research should be made on efficient segmentation techniques to enable the localization approach to be used on complex scenes.

Chapter 6
Summary and Conclusion

In this thesis three different, novel approaches are described to solve the pose estimation problem in the context of the classical bin-picking problem. All three approaches have different strengths and drawbacks and suit special requirements in different ways.

At first, in Chap. 3, a pose estimation technique, based on 3D point clouds was proposed. Here, a known technique for fragment matching (*Random Sample Matching*) was used and enhanced to work as a pose estimator. To safely interact with the scene, the localization is combined with an efficient collision avoidance mechanism. A collision measure is computed, using a gripper model positioned at a possible grasp pose in the virtual scan data. To offer a high amount of flexibility, grasp frames can be defined with certain degrees of freedom. Overall, with possible cycle times of about 12 s, and a very robust performance even in the presence of noise and clutter, the 3D point cloud based approach, built around the *Random Sample Matching* algorithm, meets typical industrial demands. Only in scenarios with extremely tight time constraints or when the cost of the system has to be very low, the approach might not be applicable. The main contributions of this chapter are:

- The description of a generic, robust and efficient pose estimation technique based on the RANSAM algorithm.
- The introduction of the *Key Grasp Frame* concept, which allows an automatic grasp planning within given ranges and therefore meets industrial demands.
- The combination of these two approaches to build a very good working bin-picking system.

To offer a solution to handle scenarios in which very tight time constraints are given, the second approach, described in Chap. 4, was developed. The Kinect was used as a 3D sensor, offering advantages regarding its low cost of only 120 US dollars and its high frame rates of 30 images per second. To benefit from the high speed measurements, it is necessary, to analyze the provided data in a very short amount of time. The reduction from 3D point clouds to 2D depth images resulted in

© Springer International Publishing Switzerland 2016

D. Buchholz, *Bin-Picking*, Studies in Systems, Decision and Control 44,

DOI 10.1007/978-3-319-26500-1_6

very efficient computations, all computations were performed on 2D depth maps. To further improve the cycle times, the sequence of the computations was changed. In the depth map based approach, the gripper poses are determined directly using the scanned depth maps. The gripper footprint is approximated as a rotation invariant 2D filter kernel. By convolution of the depth map with this kernel, graspable regions produce local maxima, which are easily extractable and directly lead to gripper pose hypotheses. These hypotheses are then analyzed for a gripper orientation, possible approach distance and a collision measure. This results in valid gripper poses in about 120 ms, including data acquisition and analysis, which means that, at every moment of execution, a valid gripper pose is available. Thus, no dead times arise and the robot can work to capacity picking up arbitrary objects. Force/torque and acceleration sensors are added to compute the moments of inertia of the grasped object which are then used to estimate the grasp pose during the robot's movements. The main contributions of this chapter are:

- A new depth map based real-time gripper pose estimation technique that can be used to grasp unknown objects and is faster and has a higher success rate than known approaches.
- The combination of this technique with a force/torque/acceleration based grasp pose estimation technique that is performed during the robot's movements and therefore does not alter the cycle times of the system.
- A depth map based bin-picking system concept that is capable of achieving the shortest possible cycle times.

   The former two pose estimation approaches are based on 3D sensors, even if the analysis is partly performed on 2D depth maps. Is bin-picking still possible, when avoiding 3D sensors? Is it possible to estimate 6D object poses, in scenarios, where laser scanners cannot be used and time of flight cameras are not accurate enough? The approach in Chap. 5 shows one way. By further reducing the dimensionality of the input data from depth maps to normal maps, a new class of optical sensors can be used as a basis for pose estimation. The described approach is the first in literature to solve generic pose estimation based on normal maps. In a first step, all normals are extracted from the image and the model and stored into spherical histograms (EGIs). By doing this, the correspondence problem is avoided. The two spheres are then efficiently correlated using the spherical fast Fourier Transform leading to an orientation estimate. Using this orientation, the translation of the object can be estimated using a 2D representation of the model or a 3D representation of the image. In combination with the orientation, this completes the 6D pose of the object relative to the camera's coordinate system. Both parts of the localization approach, the orientation and the translation estimation, can also be used without the other for further fields of application. The accuracy of the approach is comparable to 3D techniques which makes it usable in real world applications. The main contributions of this chapter are:

- A rotation estimation technique based on normal maps.
- A 2D image based technique for monocular translation estimation.

- A new 3D volume based concept for monocular translation estimation.
- The combination of these approaches for 6D pose estimation based on normal maps, which for the first time allows the usage of normal maps for generic pose estimation.

All three approaches performed very well and showed high potential w.r.t. certain aspects. But, further research has to be done in some areas.

Chapter 5 builds the main part of this thesis and describes a new way of solving the bin-picking problem or, more precisely the pose estimation problem. Therefore, not all aspects of the approach are working perfectly. At first, the computation time is quite long which can be optimized in certain ways already stated in the chapter. Further, all experiments were only done with isolated objects. Single experiments already showed the general applicability of the approach for scenes with several objects, but further work has to be done here. Two ways to tackle this open point are possible here. Either, the very global description of the normal maps using EGIs has to be changed to a local description, using for example several local EGIs of image patches, or an efficient segmentation technique to separate the objects in the image could be used. Further, the volume based translation estimation was only analyzed briefly due to the time constraints of this thesis. It showed that this technique generates the predicted results, but the quality was not as good as expected. Furthermore, the computation time was much higher than the image based method. As a closed solution is available when working with volumes, a higher precision could be possible when pursuing more research in this direction. It is interesting in this context to analyze the local neighborhood of the maxima and if the neighborhoods are usable for translation refinement. Further enhancements contain an optimization of the spherical correlation with optimizations of the computation times. Here it is notable that degrees of freedom within the orientation estimation are visible in the correlation function. This could be used not only for faster maximum search but also be exploited for self similarity analysis.

The system of Chap. 4 was only tested on quite a small scale. Further testing is needed here to show the strength of this approach. A more complex planning of estimation trajectories has to be implemented and the combination of vision and force/torque/acceleration sensing has to be further enhanced.

The approach of Chap. 3 already meets industrial demands and is ready for extensive testing. At the time of the writing of the thesis, the test platform of an industrial partner was not ready. Real world experiments and manufacturing line applicability tests are the next steps here.

These three points are topics for future work.

As a summary, three different approaches to solve the bin-picking problem are proposed within this thesis. All three techniques yield very promising results and are applicable in real world scenarios, each with different strengths and limitations.

Appendix A
Data Acquisition

There is a variety of optical sensors currently commercially available. As different types of sensors are used within the experiments of this work, a short survey of sensing techniques is presented within this chapter.

Three different types of visual data are important in this context. These types are 3D point clouds, used in Chap. 3, depth maps, used in Chap. 4 and normal maps used in Chap. 5. The first two are acquired using 3D sensors, the latter does not acquire 3D data but surface normals only. For all three data types, a variety of acquisition techniques is available. The depth sensors are very common in industry and even in consumer electronics, nowadays. Therefore, only a brief overview is given in the following section. "Normal map scanners", however, are not very common. As the normal maps play an important role within this thesis, a more detailed overview over possible acquisition techniques is given in the next section.

A.1 3D Point Cloud and Depth Map Acquisition

There is a variety of different sensors and sensor concepts to acquire 3D point clouds. Actually, any 3D sensor technique measures 3D coordinates and thus 3D point clouds. If a scanner only has one optical center and therefore only has one viewing direction, the point clouds can be stored in a 2D grid, i.e., a depth map. 3D point clouds and depth maps can be acquired using the same sensors. A review on recent sensors can be found in [18]. In industry, laser line scanners are the most common 3D scanners and there are a lot of different manufactures for this kind of sensors. Structured light scanners and time of flight cameras are further common sensors.

The main idea of triangulation based sensors, i.e. [84], is to illuminate the scene by a laser line projector. The laser line with its projection center defines a plane which is intersected by surfaces of the objects in the scene. The illuminated surface points can then easily be detected by a camera. The intersection points of the according viewing rays with the laser plane define 3D coordinates of the object's surface. The

© Springer International Publishing Switzerland 2016
D. Buchholz, *Bin-Picking*, Studies in Systems, Decision and Control 44,
DOI 10.1007/978-3-319-26500-1

advantage of this technique is that very accurate data can be acquired and dependent of the power of the laser line nearly all kinds of surfaces can be scanned. But, as only one laser line is projected, a set of images has to be taken, either by translating and/or rotating the sensor.

The same is true for structured light scanners, using for example the coded light approach, firstly described in [94]. Here, a set of stripe images is projected onto the scene and separate images are acquired. The so measured lit/unlit code in each pixel can be used to solve the correspondence problem and so, specific light planes can be found for triangulation. The spacial resolution depends on the number of projected stripe patterns which results in long acquisition times for fine scans. This approach further suffers from inter-reflections of the light patterns, where bright stripes reflect light into unlit regions of the scene, distorting the scan.

Time of flight cameras measure depth data using the known speed of light, i.e. [42, 72]. A light source emits light pulses into the scene and in every pixel of the camera the time is measured that passes until that pulse is reflected into the camera. These sensors have a limited accuracy in depth as well as spatial resolution.

A quite new and very affordable depth camera is the Microsoft Kinect sensor which is also triangulation based. An infrared projector emits a quasi random point pattern into the scene. An infrared camera detects the points and using the known pattern solves the correspondence problem enabling for triangulation. The main advantage of this sensor is its extremely low price. The Kinect is very widely used as sensor in scientific publications.

A.2 Normal Map Acquisition

Images, in which every pixel represents a surface normal are called normal maps. The classical approach to acquire normal maps is the well-known "Photometric Stereo" approach firstly proposed in [99]. Here, a set of images is taken from one camera, each using a different lighting direction. The combination of brightness values at each surface point can then be mapped to a specific surface normal. This technique will be described in more detail in Sect. A.2.2.

There are alternatives to Photometric Stereo, like Shape from Polarization, e.g. [102], Shape from Texture, e.g. [22], or Shape from Shading, e.g. [9]. Another approach to acquire normal maps uses spherical gradient illuminations to allow for multiple viewpoints and to measure specular and diffuse normal maps [61]. This approach, like the classical Photometric Stereo algorithm, needs a set of images. When cycle times of a system have to be short, a single shot characteristic of the scanning system is desirable. Therefore, the following sections describe two different approaches to acquire normal maps using single camera shots. This results in very short acquisition times and enables for a short overall cycle time.

One straight forward modification of the classical Photometric Stereo is to change the temporally multiplexed Photometric Stereo method into a spectrally multiplexed method by using colored light sources and single color images separated into their single color components, e.g. [34]. This approach will be reviewed and analyzed in this section.

Besides the spectral multiplexed Photometric Stereo, a single stripe pattern illumination technique, evaluating the local orientations and distances of the stripes in the image to get a single-shot measurement is presented in [98]. This technique overcomes issues of the multispectral Photometric Stereo but suffers from the same problems Coded Light systems suffer from, like inter-reflections. A modification of the approach of Winkelbach and Wahl [98] is presented here as well.

A.2.1 Weak Perspective Cameras

Most normal map estimation techniques use single cameras. As the estimation of the distance between an object and a camera can only be measured when a perspective camera is used, orthographic cameras cannot be applied for pose estimation techniques presented within this thesis. Using perspective cameras, the size of a surface patch observed by one pixel varies with the distance of that patch to the optical center of the camera. Here, two contrary needs arise: Because for the orientation estimation (Sect. 5.2.1), the surface area of the observed scene should be accurately sampled in the normal map which is the case with orthographic cameras and for the translation estimation (Sect. 5.2.2), a perspective camera is essential. A good compromise here is the use of a *weak perspective* camera. When the distance of the object and the camera is much larger than the thickness of the object, it can be assumed that the surface patch size observed by one pixel does not change within an object. In real scenes this is mostly the case because otherwise the camera would be too close to the scene and would interfere with the movements of the robot.

A.2.2 (Single Shot) Photometric Stereo

Photometric Stereo (PS) is a well-known technique to acquire normal maps, and using these, to generate depth data. The classical method was proposed by Woodham [99]. Many publications deal with this topic and several enhancements and modifications of the classical PS method have been developed. First modifications faced the problems of non-lambertian surfaces [23, 44]. Enhancements dealing with specularities and textured surfaces can for example be found in [21]. One of the first modifications towards single shot measurements is described in [100] where colored light sources where already used and frame rates of 15 Hz where achieved. Further modifications deal with PS using uncalibrated light sources [38], different types of light sources, like computer screens [79] or even the usage of the sun as light source ("Shape from Sun") [8].

Basic Photometric Stereo

When a surface with known reflectance properties is illuminated by a light source, the visible brightness acquired by a camera is dependent of the normal vector n of that surface. This dependence can be written as

$$E_r = E_s \cdot \mu \cdot R(s, e, n). \qquad (A.1)$$

Here, E_r denotes the light acquired by the camera, i.e. the reflected light, E_s is the incident light emitted by the light source, μ is a scalar factor named *albedo*, which describes the amount of light reflected by the surface, i.e. its "whiteness" and $R(s, e, n)$ is a reflectance function defined by the reflectance model of the surface. The reflectance function is dependent of the surface normal n, the direction of the incident light s onto the surface and the direction of the reflected light e which is the direction towards the camera.

Often, the reflectance properties of the measured surface is assumed to be diffuse, or *Lambertian*. The diffuse reflectance function is defined as

$$R_d(s, e, n) = R_d(s, n) = \begin{cases} \cos(\alpha) = \frac{s \cdot n}{|s| \cdot |n|} & \text{for } \alpha, \beta < \frac{\pi}{2} \\ 0 & otherwise \end{cases}. \qquad (A.2)$$

The diffuse reflection is independent of the direction of the reflected light. The angles α and β are the angles between the incident light s and the surface normal n, and the angle between the surface normal and the emergent light e, respectively. When the reflectance function (e.g. Lambertian reflectance, constant albedo) as well as information about the light source and the camera is known, Eq. A.1 can be solved for α. Thus, using one light source only gives information about one DOF of the surface normal. To completely reject the ambiguities, at least three images have to be taken to solve for the two DOFs (p and q) of the normals. p and q are the partial derivatives of the surface $f(x, y)$:

$$p = \frac{\delta f(x, y)}{\delta x} \qquad (A.3)$$

$$q = \frac{\delta f(x, y)}{\delta y} \qquad (A.4)$$

The normal n can then be written as

$$n = [p, q, -1]^T. \qquad (A.5)$$

To illustrate this graphically, the reflectance properties can also be represented in reflectance maps (see Fig. A.1). The detected brightness of the surface is dependent of the angle of its normal and the incident light direction. Therefore, in case of identical incident light and viewing direction, circular structures are visible in the

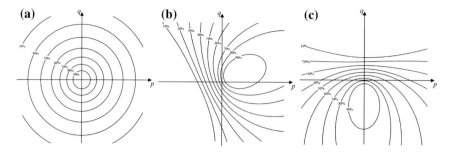

Fig. A.1 Three reflectance maps for different lighting directions

reflectance maps; on each of them the surface has the same brightness. Due to these structures, the described ambiguities arise. Three of these "iso brightness contours", under different illumination directions, intersect in exactly one point which defines one unique surface normal.

Evaluating every pixel of the three images then leads to a complete normal map.

Multi Spectral Photometric Stereo in Realistic Scenarios

To reduce the acquisition time of the classical PS method, the temporally multiplexed images can be changed into simultaneous spectrally multiplexed images. By using an RGB-Camera and three colored light sources, all images can be taken at the same time. After acquisition, the image can be separated into three color components. The Bayer-filter of the camera[1] and the spectra of the colored lights have to match, so that no light of the color channels falls into the neighboring colors. Considering the wavelength λ of the light, the reflectance equation A.1 then changes to:

$$E_r(\lambda) = E_s(\lambda) \cdot \mu(\lambda) \cdot R(s, e, n) \qquad (A.6)$$

As PS normally makes some assumptions, the usage of it in realistic scenarios has to be well planned.

Lookup Table. The evaluation of the reflectance map for each set of brightness values contains the intersection evaluation of three iso brightness contours. To reduce the computational complexity, a simple lookup table (LUT) can be sampled prior to the measurements. This can be done by using a sphere with the same reflectance properties as the measured objects and by sampling each known normal on the sphere. Each normal is then stored in a 3D table spanned by the three brightness values. A simple table lookup then replaces the intersection computation.

---

[1]A Bayer color filter is a filter array, consisting of green, red, and blue color filters, placed in front of photosensitive chips [16]. Using this filter, the gray level chips can be used to acquire color information at the cost of real resolution. Mostly, the lost resolution is regained by interpolation.

When no sphere is available, the same can be done by using arbitrary objects with known CAD models and a scan of these objects. But, the LUT might then not be filled completely.

Light Sources. It is assumed that all light sources are point light sources placed infinitely far away from the scene. This results in parallel light rays and a homogeneous brightness in the complete scene. Under realistic conditions, light sources, e.g. LEDs, have to be placed quite near to the scene. As LEDs have direction dependent varying brightness, the inhomogeneous brightness in the scene may cause errors in the measurements.

To overcome this problem, a mat white plane can be placed in the field of view of the camera. The reference brightness of each color channel of the camera can be stored and used to correct the measured brightness values later. Furthermore, the field of view can be divided into several sections, each with its own LUT, in order to evaluate pixels using only the nearest LUT.

Ambient Light. In real world situations, the measured light of the camera will be the sum of the three light sources and ambient light. Via an unilluminated reference image, the ambient light can be subtracted from the input image prior to the table lookup. This reduces errors arising from ambient light, but also reduces the dynamic range and therefore the accuracy of the measured normals.

Limitations

Photometric Stereo is well applicable using the techniques shortly introduced above. It acquires accurate and dense normal maps when the reflectance properties of the measured objects are known. Even the (colored) albedo values can be computed for each pixel.

Problems arise, when specular surfaces or surfaces with unknown reflectance properties shall be measured. Here PS is not applicable.

Further details on the method used in the experiments within this thesis can be found in [37].

A.2.3   Single Stripe Pattern Technique

In scenarios in which the PS method fails, an alternative technique has to be used. Modern laser scanners are already capable of scanning reflecting surfaces using strong lasers. Unfortunately, the acquisition times are quite high. This is because laser line triangulation sensors project a *single* laser line onto the scene, detect the reflected line in a calibrated camera image and intersect the corresponding viewing rays with the light plane to estimate 3D points. Using this procedure, one image has to be taken for each position of the laser line in the scene and all measured slices of the scene have to be merged for a complete 3D scan. To achieve a single-shot measurement, *all* laser lines have to be projected at once. Doing this, a new problem, namely the correspondence-problem, occurs. When capturing all laser lines in a single image, it is no longer trivial to decide which viewing ray intersects which light plane. Wrong correspondences (between laser lines and viewing rays) result in wrong depth values. Hence, assuming each laser light plane is attributed with an accompanying number and all planes are numbered in ascending order, it is necessary to detect all visible lines in the image and to find the correctly corresponding light plane. This problem can be solved by counting the visible lines. But, as the number of the first line in the image is unknown, the estimated numbers may contain a constant numbering error. The resulting depth values are therefore erroneous, but the relative depth changes between neighboring lines only contain very small errors. Thus, it is possible to use the depth values to generate a correct normal map by computing local gradients.

Acquisition Algorithm

In the following, it is assumed that the camera captures only one single object. If this is not the case, a segmentation algorithm has to be applied in advance.

To illuminate the scene with many light planes simultaneously, an LED projector is used, whose pose is calibrated with respect to the camera. The projected image consists of parallel lines (see Fig. A.2).

Projector and camera are arranged in such a way that the camera's field of view is fully illuminated by the projector. With the calibrated system only one image is acquired per measurement. To easily extract the light stripes in the image, red lines are projected and the red color channel of the camera is used as input image (Fig. A.4a). Then, the lines are detected with sub-pixel accuracy. Besides the line positions, the image gradients (being orthogonal to the lines) in each pixel are computed by applying the well-known Sobel operator (Fig. A.4b).

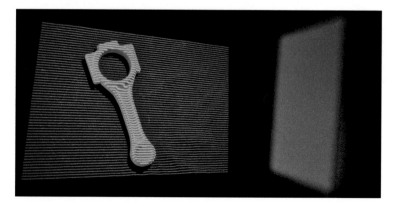

Fig. A.2 Illuminated scene from the camera's point of view of the single stripe system

To solve the correspondence problem (i.e., to assign light plane numbers to visible lines in the image), a voting algorithm is used. To prepare the algorithm, first, all pixels are clustered into single lines and each line gets an according voting table. Each voting table consists of possible line numbers with an associated weight. When the algorithm starts, all voting tables are empty. Then, each line, consisting of a set of pixels, is analyzed, starting from the top of the image (this is the border of the image that is closest to the light plane with number 0). If the voting table is empty, a number according to the position in the image is assigned depending on the average coordinate of the line pixels in the image. If the table is not empty, the current line is assigned to a light plane defined by the highest voted number. Then, for each pixel of the line, all lines in the image that intersect the normal of the line at that pixel are detected. The current pixel votes for a light plane number incremented of its own and further incremented with each intersecting line and a weight that is the inverse of the distance to the actual pixel (Fig. A.3).

When all lines are numbered, 3D coordinates can be computed for each line pixel using the estimated light plane correspondences. If a counting error occurs in the previous step, which is likely due to e.g. unknown object thickness at the topmost visible line, the estimated coordinates are erroneous. But, by applying the described voting mechanism, successive lines are assigned to successive light planes. This results in small relative depth errors between the lines.

Using the same orthogonal directions as before, for each pixel the nearest neighbors on adjacent lines are found as well as the nearest neighbors on the same line (see Fig. A.4c). All these points are used to approximate a plane and calculate its normal as surface normal in that pixel.

When all normals on each line are computed, the normal map is interpolated, again using the orthogonal directions and linearly interpolating the surface normals (Fig. A.4d).

Fig. A.3 Example of the line voting algorithm. First, line number 3 is assigned to line L_4. Then, for each pixel of L_4 all lines intersecting the orthogonal are detected and a voting is added for the incremented line number with a weight equal to the inverse of the distance of the intersection to the current pixel:

$$vote L_n \, (number, weight)$$

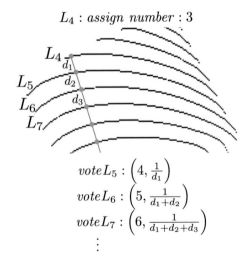

$$L_4 : assign \ number : 3$$

$$vote L_5 : \left(4, \frac{1}{d_1}\right)$$

$$vote L_6 : \left(5, \frac{1}{d_1 + d_2}\right)$$

$$vote L_7 : \left(6, \frac{1}{d_1 + d_2 + d_3}\right)$$

$$\vdots$$

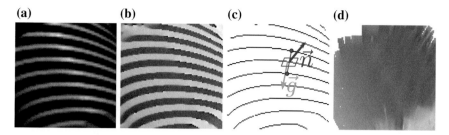

Fig. A.4 Normal map acquisition principle. **a** *Red* color channel of the input image with stripe illumination. **b** Gradient angle image. **c** Extracted lines and estimation of normal *n*, using image gradient direction. **d** Interpolated, color coded normal map

Theoretical Error Analysis

With the algorithm described in the previous section, normal maps can be generated using single camera shots. But, if the correspondences between light planes and viewing rays are erroneous, errors occur. To obtain a measure for these errors, the geometrical setup of the system can be analyzed and an error estimate can be generated. The interesting errors that occur are the depth errors f_1, f_2 of two viewing rays intersecting two adjacent light planes at the same surface as well as the normal error ϵ_N which is the angle between the measured normal and the true normal.

The errors for an example setup, depending on the numbering error ϵ_C, can be seen in Table A.1. These errors (ϵ_N and $f_{1,2}$) due to correspondence errors, depend on the extrinsic and intrinsic parameters of the camera-projector system. These parameters are d_p and d_c, the distances of the projector and the camera to the measured surface, the angle α between the camera's and the projector's direction of sight, the angle β of the projector w.r.t. the measured surface and the angle λ_{LP} between two adjacent light planes. With these parameters, ϵ_N can be computed in relation to the numbering error ϵ_C.

To get the normal error ϵ_N and the depth errors f_1 and f_2 along the viewing rays of the camera depending on the counting error ϵ_C, some geometric analysis has to be done. As the normal is measured using adjacent light planes, the angle δ between

Table A.1 Normal error ϵ_N and depth error f_1 of the single stripe pattern system depending on the correspondence error ϵ_C

ϵ_C	1	2	3	4	5
ϵ_N	0.62°	1.25°	1.89°	2.53°	3.18°
f_1	6.8 mm	13.6 mm	20.6 mm	27.7 mm	35.0 mm
ϵ_C	6	7	8	9	10
ϵ_N	3.83°	4.49°	5.16°	5.83°	6.52°
f_1	42.4 mm	49.9 mm	57.6 mm	65.5 mm	73.5 mm

Fig. A.5 Normal error ϵ_N approximation dependent of the numbering error ϵ_C

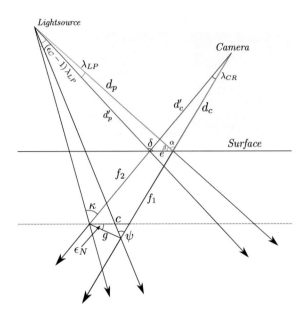

the second light plane and the surface is needed.

$$\delta = \pi - \beta - \lambda_{LP} \tag{A.7}$$

The distance e between two surface points on the surface is

$$e = d_p \frac{\sin(\lambda_{LP})}{\sin(\beta + \lambda_{LP})}. \tag{A.8}$$

Further important distances are d_p' and d_c' which are the distances of the projector and the camera along the adjacent light plane directions.

$$d_p' = d_p \frac{\sin(\beta)}{\sin(\pi - \beta - \lambda_{LP})} \tag{A.9}$$

$$d_c' = \left(d_c^2 + e^2 - 2\, d_c\, e\, \cos(\alpha + \beta)\right)^{\frac{1}{2}} \tag{A.10}$$

The angle between the viewing rays of the camera pointing towards both reflected light planes is

$$\lambda_{CR} = \sin^{-1}\left(\frac{e}{d_c'} \sin(\alpha + \beta)\right). \tag{A.11}$$

To calculate the depth error, two further angles, κ and ψ, are needed.

$$\kappa = \alpha + \lambda_{CR} - (\epsilon_C + 1)\,\lambda_{LP} \tag{A.12}$$

$$\psi = \alpha - \epsilon_C\,\lambda_{LP} \tag{A.13}$$

Now f_1 and f_2 can be estimated.

$$f_1 = d_p \frac{\sin(\epsilon_C\,\lambda_{LP})}{\sin(\psi)} \tag{A.14}$$

$$f_2 = d_p{}' \frac{\sin(\epsilon_C\,\lambda_{LP})}{\sin(\kappa)} \tag{A.15}$$

The normal error ϵ_N then is

$$\epsilon_N = \sin^{-1}\left(\frac{(f_1 - f_2)}{g}\,\sin(\pi - \alpha - \beta)\right), \tag{A.16}$$

with

$$g = \left((d_c + f_1)^2 + \left(d_c{}' + f_2\right)^2 - 2\,(d_c + f_1)\,\left(d_c{}' + f_2\right)\,\cos(\lambda_{CR})\right)^{\frac{1}{2}}. \tag{A.17}$$

In the experimental setup, the parameters were $d_p = 650\,\text{mm}$, $d_c = 471\,\text{mm}$, $\alpha = 30°$, $\beta = 69°$ and $\lambda_{LP} = 0.3°$. The resulting theoretical errors can be seen in Table A.1. The displayed error f_1 is just stated for reasons of completeness as it has no effect on the normal map.

Own Publications and References

Own Publications

1. D. Buchholz, M. Futterlieb, S. Winkelbach, and F. Wahl. Efficient bin-picking and grasp planning based on depth data. In *Proc. of International Conference on Robotics and Automation (ICRA)*, pages 3230–3235, 2013
2. D. Buchholz, D. Kubus, I. Weidauer, A. Scholz, and F. Wahl. Combining visual and inertial features for efficient grasping and bin-picking. In *Proc. of International Conference on Robotics and Automation (ICRA)*, pages 875–882, 2014
3. D. Buchholz, D. Kubus, S. Winkelbach, and F. Wahl. 3d object localization using single camera images. In Proc. of International Conference on Pattern Recognition (ICPR), pages 821–824, 2012
4. D. Buchholz, S. Winkelbach, and F. Wahl. Ransam for industrial bin-picking. In Proc. of International Symposium on Robotics (ISR) / Robotik, pages 1317–1322, 2010
5. F. Dietrich, D. Buchholz, F. Wobbe, F. Sowinski, A. Raatz, W. Schumacher, and F. Wahl. On contact models for assembly tasks: Experimental investigation beyond the peg-in-hole problem on the example of force-torque maps. In Proc. of International Conference on Intelligent Robots and Systems (IROS), pages 2313–2118, 2010
6. F. Dietrich, F. Wobbe, D. Buchholz, F. Sowinski, A. Raatz, W. Schumacher, and F. Wahl. Enhancements of force-torque map based assembly applied to parallel robots. In Proc. of International Conference on Industrial Technology (ICIT), pages 469–474, 2010
7. S. Winkelbach, J. Spehr, D. Buchholz, M. Rilk, F. Wahl, Shape (self-)similarity and dissimilarity rating for segmentation andmatching. In *Proc. of Pattern Recognition (DAGM-OAGM). Lecture Notes in Computer Science*, volume 7476, pages 93–102, 2012

References

8. A. Abrams, C. Hawley, R. Pless, Heliometric stereo: shape from sun position, in *Proceedings of the European Conference on Computer Vision*, Lecture Notes in Computer Science (2012), pp. 357–370
9. A.H. Ahmed, Shape from shading for hybrid surfaces, in *Proceedings of IEEE International Conference on Image Processing* (2007), pp. 525–528

© Springer International Publishing Switzerland 2016
D. Buchholz, *Bin-Picking*, Studies in Systems, Decision and Control 44,
DOI 10.1007/978-3-319-26500-1

10. A. Alba, R. Aguilar-Ponce, J. Vigueras-Gómez, E. Arce-Santana, Phase correlation based image alignment with subpixel accuracy, in *Advances in Artificial Intelligence*, ed. by I. Batyrshin, M. González Mendoza. Lecture Notes in Computer Science, vol. 7629 (Springer, Heidelberg, 2013), pp. 171–182

11. V. Argyriou, T. Vlachos, Sub-pixel motion estimation using gradient cross-correlation, in *Proceedings of Seventh International Symposium on Signal Processing and Its Applications*, vol. 2 (2003), pp. 215–218

12. Aristotle, *Politics—A Treatise on Government*. 350 BC

13. M.L. Baird, Image segmentation technique for locating automotive parts on belt conveyors, in *Proceedings of IJCAI* (1977), pp. 694–695

14. D. Ballard, Generalizing the hough transform to detect arbitrary shapes. Pattern Recognit. **13**, 111–122 (1981)

15. J. Baumgartl, D. Henrich, Fast vision-based grasp and delivery planning for unknown objects, in *Proceedings of German Conference on Robotics* (2012), pp. 1–5

16. B.E. Bayer, Color imaging array (1976) US Patent 3,971,065

17. M. Berger, G. Bachler, S. Scherer, Vision guided bin picking and mounting in a flexible assembly cell, in *Intelligent Problem Solving. Methodologies and Approaches. Lecture Notes in Computer Science*, vol. 1821 (2000), pp. 255–321

18. F. Blais, Review of 20 years of range sensor development. J. Electron. Imaging **13**(1), 231–240 (2004)

19. R.C. Bolles, R.A. Cain, Recognizing and locating partially visible objects: the local-feature-focus method. Int. J. Robot. Res. **1**(3), 57–82 (1982)

20. R.C. Bolles, R.P. Horaud, 3DPO: a three dimensional part orientation system. Int. J. Robot. Res. **5**(3), 3–26 (1986)

21. C. Cho, H. Minamitani, A new photometric method using 3 point light sources. IEICE Trans. Inf. Syst. **76**(8), 898–904 (1993)

22. M. Clerc, S. Mallat, The texture gradient equation for recovering shape from texture. IEEE Trans. Pattern Anal. Mach. Intell. **24**(4), 536–549 (2002)

23. E.N.J. Coleman, R. Jain, Obtaining 3-dimensional shape of textured and specular surfaces using four-source photometry. Comput. Graph. Image Process. **18**(4), 309–328 (1982)

24. J.J. Craig, *Introduction to Robotics: Mechanics and Control* (Prentice Hall, 2003)

25. J.-D. Dessimoz, J.R. Birk, R.B. Kelley, H.A.S. Martins, C. L. I. Matched filters for bin picking. IEEE Trans. Pattern Anal. Mach. Intell. **6**(6), 686–697 (1984)

26. P. Diaconis, M. Shahshahani, The subgroup algorithm for generating uniform random variables. Probab. Eng. Inf. Sci. **1**, 15–32 (1987)

27. D. Fiedler, H. Müller, Impact of thermal and environmental conditions on the kinect sensor, in *Proceedings of International Workshop on Depth Image Analysis at the 21st International Conference on Pattern Recognition* (2012)

28. B. Finkemeyer, T. Kröger, D. Kubus, M. Olschewski, F.M. Wahl, MiRPA: Middleware for robotic and process control applications, in *Workshop on Measures and Procedures for the Evaluation of Robot Architectures and Middleware at the IEEE/RSJ International Conference on Intelligent Robots and Systems* (2007), pp. 78–93

29. D. Fischinger, M. Vincze, Learning grasps for unknown objects in cluttered scenes, in *Proceedings of IEEE International Conference on Robotics and Automation* (2013), pp. 601–608

30. O. Fischler, R. Bolles, Random sample consensus: a paradigm for model fitting with applications to image analysis and automated cartography. *Communications of the Association for Computing Machinery* (1981), pp. 381–395

31. A.J. Fitch, A. Kadyrov, W.J. Christmas, J. Kittler, Orientation correlation, in *Proceedings of British Machine Vision Conference* (2002), pp. 133–142

32. G. Forsen, *Processing Visual Data with an Automation Eye*. Pictorial Pattern Recognition (Thompson Book Co., Washington D.C., 1968)

33. J.H. Friedman, J.L. Bentley, R.A. Finkel, An algorithm for finding best matches in logarithmic expected time. ACM Trans. Math. Softw. **3**, 209–226 (1977)

34. G. Fyffe, X. Yu, P. Debevec, Single-shot photometric stereo by spectral multiplexing, in *Proceedings of IEEE International Conference on Computational Photography* (2010), pp. 1–6

35. O. Ghita, P.F. Whelan, A bin picking system based on depth from defocus. *Machine Vision and Applications* (2003), pp. 234–244

36. A. Gonzalez, Measurement of areas on a sphere using fibonacci and latitude-longitude lattices. Math. Geosci. **42**(1), 49–64 (2010)

37. E. Günter, Erzeugung von oberflächennormalen-karten unter realistischen bedingungen aus einem einzelnen kamerabild. Master's thesis, Technische Universität Braunschweig (2014)

38. H. Hayakawa, Photometric stereo under a light source with arbitrary motion. J. Opt. Soc. Am. **11**(11), 3079–3089 (1994)

39. B.K.P. Horn, Extended gaussian images, in *Proceedings of the IEEE* (1984), pp. 1671–1686

40. B.K.P. Horn, K. Ikeuchi, The mechanical manipulation of randomly oriented parts. Sci. Am. **251**(2), 100–111 (1984)

41. B. Huhle, T. Schairer, W. Strasser, Normalized cross-correlation using soft, in *International Workshop on Local and Non-Local Approximation in Image Processing* (2009), pp. 82–86

42. S. Hussmann, T. Ringbeck, B. Hagebeuker, A performance review of 3D TOF vision systems in comparison to stereo vision systems. *Stereo Vision* (2008), pp. 103–120

43. D.Q. Huynh, Metrics for 3D rotations: comparison and analysis. J. Math. Imaging Vis. **35**(2), 155–164 (2009)

44. K. Ikeuchi, Determining surface orientations of specular surfaces by using the photometric stereo method. IEEE Trans. Pattern Anal. Mach. Intell. **3**(6), 661–669 (1981)

45. K. Ikeuchi, B.K. Horn, S. Nagata, Tom, T. Callahan, O. Feingold, Picking up an object from a pile of objects, in *Proceedings of the First International Symposium on Robotics Research.* MIT Press (1983), pp. 139–166

46. R. Iser, D. Kubus, F.M. Wahl, An efficient parallel approach to random sample matching (pransam), in *Proceedings of IEEE International Conference on Robotics and Automation* (2009), pp. 1199–1206

47. S. Julier, The scaled unscented transformation. Proc. Am. Control Conf. **6**, 4555–4559 (2002)

48. M. Kazhdan, T. Funkhouser, S. Rusinkiewicz, Rotation invariant spherical harmonic representation of 3D shape descriptors, in *Proceedings of the 2003 Eurographics/ACM SIGGRAPH Symposium on Geometry Processing* (2003), pp. 156–164

49. J. Keiner, D. Potts, Fast evaluation of quadrature formulae on the sphere. Math. Comput. **77**, 397–419 (2008)

50. Z. Khalid, S. Durrani, R.A. Kennedy, P. Sadeghi, On the construction of low-pass filters on the unit sphere, in *Proceedings of International Conference on Acoustics, Speech and Signal Processing* (2011), pp. 4356–4359

51. M. Kolomenkin, I. Shimshoni, A. Tal, On edge detection on surfaces, in *Proceedings of IEEE Conference on Computer Vision and Pattern Recognition* (2009), pp. 2767–2774

52. P.J. Kostelec, D.N. Rockmore, FFTs on the rotation group. J. Fourier Anal. Appl. **14**(2), 145–179 (2008)

53. D. Kubus, T. Kröger, F. Wahl, On-line rigid object recognition and pose estimation based on inertial parameters, in *Proceedings of International Conference on Intelligent Robotics and System* (2007), pp. 1402–1408

54. D. Kubus, T. Kröger, F. Wahl, On-line estimation of inertial parameters using a recursive total least-squares approach, in *Proceedings of International Conference on Intelligent Robotics and System* (2008), pp. 3845–3852

55. D. Kubus, A. Sommerkorn, T. Kröger, J. Maass, F. Wahl, Low-level control of robot manipulators: distributed open real-time control architectures for Stäubli RX and TX manipulators, in *Workshop on Innovative Robot Control Architectures for Demanding (Research) Applications—How to Modify and Enhance Commercial Controllers at the IEEE International Conference on Robotics and Automation* (2010), pp. 38–45

56. A. Kudla, Korrelationsbasierte posenschätzung. Master's thesis, Technische Universität Braunschweig (2014)

57. Y. Lamdan, H. Wolfson, Geometric hashing: a general and efficient model-based recognition scheme, in *Proceedings of International Conference on Computer Vision* (1988), pp. 238–249

58. R.L. Larkins, M.J. Cree, A.A. Dorrington, Analysis of binning of normals for spherical harmonic cross-correlation, in *Proceedings of Three-Dimensional Image Processing (3DIP) and Applications*, vol. 8290 (2012)
59. L. Ljung, *System Identification: Theory for the User* (Prentice Hall, 1999)
60. D.G. Lowe, Distinctive image features from scale-invariant keypoints. Int. J. Comput. Vis. **60**, 91–110 (2004)
61. W.-C. Ma, T. Hawkins, P. Peers, C.-F. Chabert, M. Weiss, P. Debevec, Rapid acquisition of specular and diffuse normal maps from polarized spherical gradient illumination, in *Proceedings of Eurographics Symposium on Rendering* (2007), pp. 183–194
62. A. Makadia, A. Patterson, K. Daniilidis, Fully automatic registration of 3D point clouds, in *Proceedings of the IEEE Computer Society Conference on Computer Vision and Pattern Recognition* (2006), pp. 1297–1304
63. A. Mian, M. Bennamoun, R. Owens, On the repeatability and quality of keypoints for local feature-based 3D object retrieval from cluttered scenes. Int. J. Comput. Vis. **89**(2–3), 348–361 (2010)
64. A.S. Mian, M. Bennamoun, R. Owens, Three-dimensional model-based object recognition and segmentation in cluttered scenes. IEEE Trans. Pattern Anal. Mach. Intell. **28**(10), 1584–1601 (2006)
65. A. Miller, S. Knoop, H. Christensen, P. Allen, Automatic grasp planning using shape primitives, in *Proceedings of IEEE International Conference on Robotics and Automation*, vol. 2 (2003), pp. 1824–1829
66. J.C. Mitchell, Sampling rotation groups by successive orthogonal images. SIAM J. Sci. Comput. **30**(1), 525–547 (2008)
67. J.-K. Oh, K. Baek, D. Kim, S. Lee, Development of structured light based bin picking system using primitive models, in *Proceedings of IEEE International Symposium on Assembly and Manufacturing* (2009), pp. 46–52
68. C. Papazov, D. Burschka, An efficient ransac for 3D object recognition in noisy and occluded scenes, in *Proceedings of Asian Conference on Computer Vision* (2011), pp. 135–148
69. C. Papazov, S. Haddadin, S. Parusel, K. Krieger, D. Burschka, Rigid 3D geometry matching for grasping of known objects in cluttered scenes. Int. J. Robot. Res. **31**(4), 538–553 (2012)
70. W.A. Perkins, A model-based vision system for industrial parts. IEEE Trans. Comput. **C-27**(2), 126–143 (1978)
71. PLM Engineering: Betriebsautomation (2013), http://plm-engineering.com/leistungen/betriebsautomation/index.php
72. T. Ringbeck, B. Hagebeuker, A 3D time of flight camera for object detection, in *Proceedings of Conference On Optical 3-D Measurement Techniques* (2007)
73. L.G. Roberts, Machine Perception of Three-Dimensional Solids. PhD thesis, Massachusetts Institute of Technology (1963)
74. F. Röhrdanz, Modellbasierte automatisierte Greifplanung. PhD thesis, Institut für Robotik und Prozessinformatik, Technnische Universität Braunschweig (1998)
75. F. Röhrdanz, R. Gutsche, F.M. Wahl, Assembly planning and geometric reasoning for grasping, in *Proceedings of International Conference on Artificial Intelligence and Information-Control Systems of Robots* (1994), pp. 93–106
76. F. Röhrdanz, F.M. Wahl, Generating and evaluating regrasp operations, in *Proceedings of International Conference on Robotics and Automation* (1997), pp. 2013–2018
77. K. Safronov, I. Tchouchenkov, H. Wörn, Hierarchical iterative pattern recognition method for solving bin picking problem, in *Proceedings of Robotik* (2008), pp. 3–6
78. A. Saxena, J. Driemeyer, A.Y. Ng, Robotic grasping of novel objects using vision. Int. J. Robot. Res. **27**(2), 157–173 (2008)
79. G. Schindler, Photometric stereo via computer screen lighting for real-time surface reconstruction, in *Proceedings of the Fourth International Symposium on 3D Data Processing, Visualization and Transmission* (2008)
80. A. Schyja, A. Hypki, B. Kuhlenkotter, A modular and extensible framework for real and virtual bin-picking environments, in *Proceedings of International Conference on Robotics and Automation* (2012), pp. 5246–5251

81. Y. Shirai, M. Suwa, Recognition of polyhedrons with a range finder, in *Proceedings of the 2nd International Joint Conference on Artificial Intelligence* (1971), pp. 80–87

82. SICK: Data sheet sick LMS 400-1000 (2013), https://www.mysick.com/partnerPortal/ProductCatalog/DataSheet.aspx?ProductID=33774

83. SICK: Data sheet sick LMS 500-20000 (2014), https://www.mysick.com/partnerPortal/ProductCatalog/DataSheet.aspx?ProductID=45446

84. SICK IVP: Data sheet sick IVP ruler e1200 (2013), https://www.mysick.com/partnerPortal/ProductCatalog/DataSheet.aspx?ProductID=37310

85. L. Sorgi, K. Daniilidis, Normalized cross-correlation for spherical images, in *Proceedings of European Conference on Computer Vision*, vol. 3022 (2004), pp. 542–553

86. T. Stahs, Ein aktives 3D-Robotersensorsystem auf der Grundlage eines verallgemeinerten Ansatzes zur Erstellung modellbasierter Objekterkennungsverfahren. Ph.D. thesis, Institut für Robotik und Prozessinformatik, Technische Universität Braunschweig (1994)

87. R. Swinbank, J.R. Purser, Fibonacci grids: a novel approach to global modelling. Q. J. R. Meteorol. Soc. **132**(619), 1769–1793 (2006)

88. S. Tsuji, A. Nakamura, Recognition of an object in a stack of industrial parts, in *Proceedings of 4th IJCAI: 8 1 1* (1975), pp. 811–818

89. J. Turney, T. Mudge, R. Volz, Recognizing partially hidden objects. Proc. IEEE Int. Conf. Robot. Autom. **2**, 48–54 (1985)

90. G. Tzimiropoulos, V. Argyriou, S. Zafeiriou, T. Stathaki, Robust FFT-based scale-invariant image registration with image gradients. IEEE Trans. Pattern Anal. Mach. Intell. **32**, 1899–1906 (2010)

91. J. van den Berg, P. Abbeel, K. Goldberg, LQG-MP: optimized path planning for robots with motion uncertainty and imperfect state information. Int. J. Robot. **30**(7), 895–913 (2011)

92. P. Viola, M. Jones, Rapid object detection using a boosted cascade of simple features. In *Proceedings of IEEE Computer Society Conference on Computer Vision and Pattern Recognition*, vol. 1 (2001), pp. 511–518

93. A. Vollrath, The nonequispaced fast SO(3) fourier transform, generalisations and applications. Ph.D. thesis, Institut für Mathematik, Universität Lübeck (2010)

94. F.M. Wahl, A coded light approach for depth map acquisition. Informatik-Fachberichte **125**, 12–17 (1986)

95. E.W. Weisstein, Birthday attack. From Math World—A Wolfram Web Resource (2008), http://mathworld.wolfram.com/BirthdayAttack.html

96. S. Winkelbach, Das 3D-Puzzle-Problem—Effiziente Methoden zum paarweisen Zusammensetzen von dreidimensionalen Fragmenten. Ph.D. thesis, Institut für Robotik und Prozessinformatik, Technische Universität Braunschweig (2006)

97. S. Winkelbach, S. Molkenstruck, F.M. Wahl, Low-cost laser range scanner and fast surface registration approach. Pattern Recognit.—Lect. Notes Comput. Sci. **4174**, 718–728 (2006)

98. S. Winkelbach, F.M. Wahl, Shape from single stripe pattern illumination. Pattern Recognit.—Lect. Notes Comput. Sci. **2449**, 240–247 (2002)

99. R.J. Woodham, Photometric method for determining surface orientation from multiple images. Opt. Eng. **19**, 139–144 (1980)

100. R.J. Woodham, Gradient and curvature from photometric stereo including local confidence estimation. J. Opt. Soc. Am. **11**, 3050–3068 (1994)

101. A. Zaharescu, E. Boyer, K. Varanasi, R. Horaud, Surface feature detection and description with applications to mesh matching, in *Proceedings of IEEE Conference on Computer Vision and Pattern Recognition* (2009), pp. 373–380

102. L. Zhang, E. Hancock, Simultaneous reflectance estimation and surface shape recovery using polarisation, in *Proceedings of International Conference on Pattern Recognition (ICPR)* (2012), pp. 1876–1879

103. S. Zhou, R. Chellappa, From sample similarity to ensemble similarity: probabilistic distance measures in reproducing kernel hilbert space. IEEE Trans. Pattern Anal. Mach. Intell. **28**(6), 917–929 (2006)